Mobile Radio Servicing Handbook

SUPPLIED BY

Directorate of Telecommunications
Communications Training Centre
B.Blk., Government Buildings
London Road
Stanmore
Middlesex. HA7 4QE.

Tel. 081 958 8511

Home Office Communications Training Centre

Mobile Radio Servicing Handbook

Roger Belcher, Mike Fitch, David Ogley and Geoff Varrall

Heinemann Newnes

Heinemann Newnes
An imprint of Heinemann Professional Publishing Ltd
Halley Court, Jordan Hill, Oxford OX2 8EJ

OXFORD LONDON MELBOURNE AUCKLAND SINGAPORE
IBADAN NAIROBI GABORONE KINGSTON

First published 1989
© Roger Belcher, Mike Fitch, David Ogley and Geoff Varrall 1989

British Library Cataloguing in Publication Data
Mobile radio servicing handbook.
  1. Mobile radio systems
  I. Belcher, Roger
  621.3841'65

ISBN 0 434 92187 4

Typeset by Deltatype, Ellesmere Port
Printed and bound in Great Britain by
Billings Ltd, Worcester.

Contents

Preface	xi
Acknowledgements	xiii
About the authors	xv

1 Mobile radio in perspective — 1
  New component technology – new production processes — 4
  Impact on workshop repair procedures — 5
  New technology across the frequency spectrum — 6
  New modulation techniques — 7
  Data over radio — 7
  Digital speech systems — 8
  Impact of innovation — 9

2 Propagation and frequency utilization — 11
  Frequency and wavelength — 11
  Standing waves — 12
  Spectrum utilization — 14
  Spectrum propagation characteristics — 15
  Frequency allocation — 17
  Radio frequency – spectrum allocation — 20

3 System choices — 21
  Cellular radio — 21
  System 4 — 21
  Wide area paging — 22
  On-site paging — 22
  Private mobile radio (PMR) — 23
  CT2 'phone zone' or 'telepoint' — 24
  Pan-European (Group Special Mobile) digital cellular system — 24
  Service and repair requirements for each system — 25

Contents

4 Principles of RF communication — 26
Introduction — 26
Concept of modulation — 26
Amplitude modulation (AM) — 27
Frequency modulation (FM) — 29
Phase modulation (PM) — 32
Pulse code modulation (PCM) — 33
Receiver fundamentals — 34
Transmitter fundamentals — 36

5 Practice of RF communication — 40
Introduction — 40
Antennas and matching — 40
RF amplification — 41
Mixing — 44
IF filtering — 46
IF amplification and detection for AM — 48
IF amplification and detection for FM — 49
Squelch and audio output — 51

6 Practice of RF communication – transmitter design — 53
Audio processing — 53
RF signal generation — 56
Modulating the RF signal — 57
Amplification — 58

7 Frequency synthesizer circuit principles — 60
Introduction — 60
RF scillator – theory and practice — 60
Frequency and phase comparators — 67
The loop filter — 68
The phase locked loop PLL — 70
Dual modulus prescaling — 74
Practical considerations — 82

8 Principles of RF measurement
Test and measurement objectives — 84
RF fundamentals — 84
Intermodulation distortion — 84
Time and frequency analysis — 88
Use of decibels as a measurement base — 93
Receiver parameter definitions — 94
Transmitter parameter definitions — 95

	Receiver tests	97
	Transmitter tests	105
9	**Calibration, test and fault finding**	111
	Test equipment requirements and use	111
	Alignment and calibration	114
	In-service fault finding – receivers	125
	In-service fault finding – transmitters	137
10	**Component identification and handling**	145
	Classification of components	145
	Discrete semiconductor device applications	152
	Integrated circuits	161
	Passive components	161
	Handling semiconductors	164
	PCB track repair	167
	Surface mount technology	167
11	**Electromagnetic compatibility (EMC)**	173
	Introduction	173
	Methods of reducing radiation	173
	Methods of reducing sensitivity to radiation	173
	Installation techniques to minimize interference	176
	Sources of radio interference	179
12	**Antennas – selection, installation, fault finding and maintenance**	181
	Introduction	181
	Antenna gain and patterns	181
	Directional antennas	183
	Stacked and bayed yagis	186
	Other styles of directional antennas	188
	Omni directional antennas	191
	Beam tilt	195
	Baying omni directional antennas	196
	Cardiod Dipole	197
	Choosing omni directional antennas	198
	The polo mint effect	199
	Wide band omni directional antennas	200
	Various other styles and applications for antennas	201
	Specification of fixed antennas	203
	Assessing an antenna's performance	205
	Choice of radio site	206
	Installation of fixed antennas	207

	Fault diagnosis	211
	Environmental problems	212
	Health and safety	213
	Duplexers and filters	214
	Dish antennas for microwave and millimetric wavelengths	217
	Link planning	218
13	**Typical operator and system problems**	222
	Test calls	222
	Batteries	223
	Health and safety	224
	Service and support philosophy	224
14	**Spectrum efficiency – audio Selcall, trunking and cellular systems**	226
	Strategies for spectrum efficiency	226
	Audio Selcall (selective calling)	226
	Digital signalling using FFSK	230
	Trunking	232
	Cellular radio	237
	UK TACS structure – spectrum utilization	240
	Implications for testing and in-service support	240
	Summary	243
15	**Pan-European digital technology**	244
	Digital speech and digital modulation/processing techniques	244
	Speech synthesis	244
	Signal preparation	245
	Speech quality	245
	Modulation techniques	246
	Transmission techniques	246
	Signalling control protocols	247
	Channel loading implications	247
	CT2 phone zone cordless telephone applications	248
	Summary	249
	The future	249
16	**Data over radio**	251
	Introduction	251
	ASCII 'strings'	251
	Parity checks	252
	Data over cellular	253
	Shift keying	253
	Bandwidth	255
	Channel capacity	255

Appendix I	*Abbreviations and glossary of terms*	257
Appendix II	*Terminology*	259
Appendix III	*The frequency spectrum*	264
Appendix IV	*Abbreviations for mobile radio users*	266
Appendix V	*Private mobile radio bands*	268
Appendix VI	*Conversion tables*	271
Appendix VII	*Mobile communications and ISDN (Integrated Services Digital Network)*	275
Appendix VIII	*Sources of help and advice*	277
Index		279

Preface

The objective of this handbook is to provide a useful, practical package of information on the servicing and repair of VHF and UHF mobile radios and base stations, together with the maintenance and support requirements of the overall radio system, including antenna and mast installations.

The fast changing component and system technology of mobile radio communications is having a major impact on the way in which equipment is installed, commissioned, serviced and maintained. The 'Mobile Radio Servicing Handbook' sets out to provide the technician or engineer with sufficient information to be able to undertake, or plan, repair and maintenance work on a modern mobile radio system, with hand held or vehicle mounted mobiles. In addition we look at the impact of data over radio, cellular radio, and trunking technologies on servicing, diagnosis and repair procedures.

Topics covered as background include radio theory, amplitude (AM) and frequency (FM) modulation, radio wave propagation, reception and demodulation, fundamentals of receiver and transmitter systems, principles of transmitter and receiver design, synthesizer techniques, selective tone signalling and the digital signalling control and access protocols of cellular and trunked radio, but the Handbook is intended primarily as a source of readily accessible information on how to fault find and repair mobile radio equipment – mobiles, base stations and antennas – to circuit module and component level.

Detailed appendices include information on areas of topical interest, including data over radio, frequency allocation, sources of help and advice, and day to day reference information.

We have very much enjoyed putting the Handbook together and hope you find it a useful and authoritative companion at the workbench.

Acknowledgements

The authors would like to offer their special thanks to the following for their valued help and advice: Don Baker, General Manager, Aerial Sites Ltd, Chesham, Bucks; Terry Platt and Jack Varrall for their patient proof reading and contributions to the text. Elizabeth Gossage for her heroic work on the typescript. Historical photographs courtesy of Philips Radio Communication Systems, Cambridge. Antenna illustrations courtesy of Jaybeam Antennas, Northampton. Transceiver photographs courtesy of Tait Mobile Radio, Huntingdon.

About the authors

Roger Belcher

Roger Belcher has a background of twenty five years in the RF communications industry, including an apprenticeship at Marconi, design and development of antennae and radar processing equipment for Plessey Radar (1967–72), technical marketing for Texas Instruments (1972–77), antenna and modem design for satellite communications for Racal (1977–80) and design of RF test equipment and associated hardware/software development for Rohde and Schwarz (1980–84). He is Technical Director of the test equipment and training company RTT.

Mike Fitch

Michael Fitch worked for Marconi Space and Defence Systems for six years as Project Manager for satellite ground stations, followed by three years at Tait Electronics, New Zealand as Senior Design Engineer responsible for base station products and six years managing the technical support services for Tait Mobile Radio in the U.K. He is currently an executive engineer with British Telecom; Martlesham Heath, managing a small team developing base station products for Pan-European digital mobile radio.

David Ogley

After a career in the Royal Signals as a radio engineer, David Ogley has worked in the mobile radio industry for the past seven years and is currently undertaking survey and design work for Jaybeam Antennas in the UK as General Manager of Jaybeam Installations Ltd. Division.

Geoff Varrall

Geoff Varrall graduated from St John's College, Cambridge in the mid 1970s and became well known as a training and educational specialist. He has developed a number of training courses for contractors and installers in the electrical and electronic industries, and since working with RTT has been responsible for the joint RF technology training courses organized in association with Jaybeam Antennas and Aerial Facilities Ltd.

1 Mobile radio in perspective

Over the next few chapters, we will be focusing on UHF/VHF mobile radio–transceiver circuit design, RF operational parameters, antenna systems, general fault-finding and diagnostic routines – looking at how the introduction of new technology at component, circuit and system level is influencing operational parameters, maintenance and repair procedures.

Today's mobile radio is very different from the radio of the 1960s and 1970s. We have to adapt to the process of change in the same way that our predecessors had to adapt to the challenge of technical innovation.

Each major step forward in radio technology has required a parallel shift in the way in which radios are managed and maintained. From the experiments of Hughes and Hertz in the 1880s, we can trace the impact of change through one hundred years of evolution:

- Rutherford's first wireless transmission from the Cavendish Laboratory in Cambridge in 1896.
- Fessenden's attempts at wireless telephony in 1900 using high speed spark generation.
- The development of crude, but powerful, spark transmitter technology used by Marconi to make the first trans-Atlantic radio transmission in 1901.
- Fleming's thermionic valve in 1904 providing rectification, and the future basis for radio telephony (speech over radio) as opposed to radio telegraphy (morse) applications.
- Lee de Forest's triode valve in 1907 (detector, amplifier and generator).
- The development and application of tuned circuits by the Marconi Company in the First World War (the basis for 'frequency channels', giving longer distance, lower power communication capability).
- The application of portable morse code transceivers in the First World War.
- The development of quality long distance radio transmission for commercial applications in the 1920s and 1930s
  1 the formation of the BBC in 1922;
  2 the Hague broadcasts in the early 1920s – the first successful broadcasting of music over radio;
  3 the progressive development of short wave, long distance transmissions in the 1920s to the Commonwealth (including the pioneering work done by radio amateurs such as Gerald Marcuse in the UK);

4 George V addressing the Empire in 1932 (on the short wave world service).
- The introduction by Major Edwin Armstrong of FM modulation into radio broadcasting in the USA in 1939.
- The development of 'portable' VHF and short wave two-way radios in the Second World War using new mass production techniques (over 40 000 infantry radios were manufactured by Pye in one production run).

Figure 1.1 *Pye WS62 short wave mobile radio set used in the Second World War*

- The introduction of FM VHF mobiles – first in the US and then in the UK towards the end of the War.
- The introduction of the first land mobile radio service in Cambridge in 1947 using wartime production and component technology – the true beginning of the UK land mobile industry.
- The use of two-way VHF radio in exploration.

The years following the Second World War have seen perhaps the most impressive year-on-year advances in performance, for example:

- A rapid increase in selectivity and frequency stability with a reduction in practical channel widths from 180–200 kHz in 1945 to 12.5 kHz/6.25 kHz today.
- Dramatic reductions in power demand, particularly following the introduction of printed circuit boards and transistorisation in the late 1950s and early 1960s with power requirements reducing from 100 watts to 1–2 watts for receivers.
- Dramatic reductions in weight – 40 kg for 25 watt receivers down to 1–2 kg today.

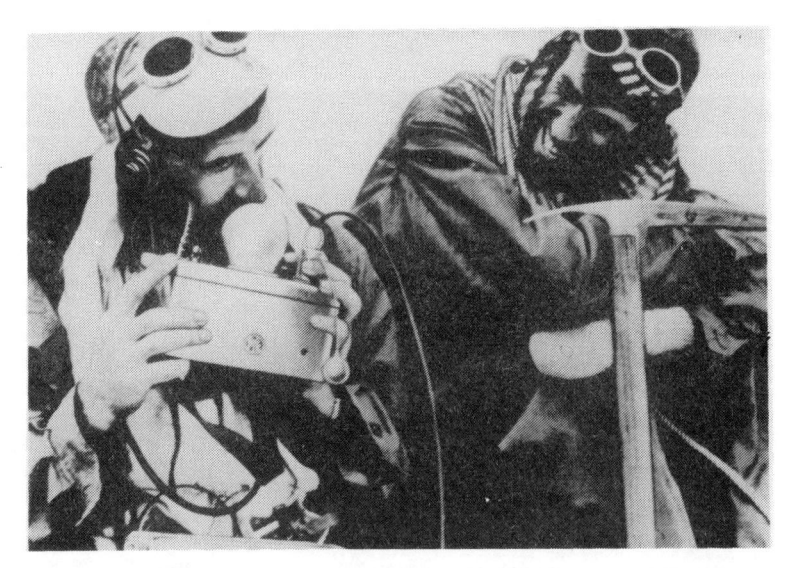

Figure 1.2 *Pye VHF two-way radio being used in Sir John Hunt's expedition to Everest in 1953*

- Dramatic reductions in size, from shoulder pack to pocket book, leading to new applications, for instance the development of UHF portable radios for the police in the early 1960s.
- Improvements in passive components, for example:
  1 the use of new dialectric materials such as plastics, ceramics and titanium electrolytes to reduce capacitor dimensions;
  2 the use of improved ferrites for smaller higher stability inductors;
  3 the use of helical resonators, strip lines and SAW devices to improve tuned circuits;
  4 the etching of conductor and resistor patterns on ceramic or glass substrates using thin and thick film technology to increase surface utilization;
  5 the introduction of dot matrix and alpha numeric liquid crystal displays.
- Improvements in active components. The thermionic valve remained the most effective amplifying device until well into the 1950s (new miniaturized valves were still being introduced by RCA as late as 1960), but with the change from germanium to silicon, the performance of transistor devices increased rapidly. Today a single transistor power amplifier device can deliver up to 40 watts at 900 MHz.
- Development of circuit design: The development of synthesizer circuits making possible the introduction of 'intelligent' frequency hopping radios and the implementation of cellular and trunked radio networks.

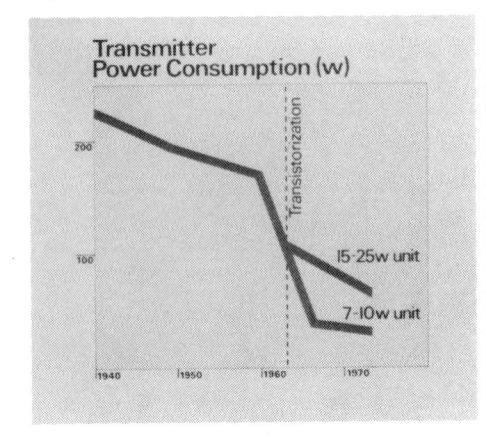

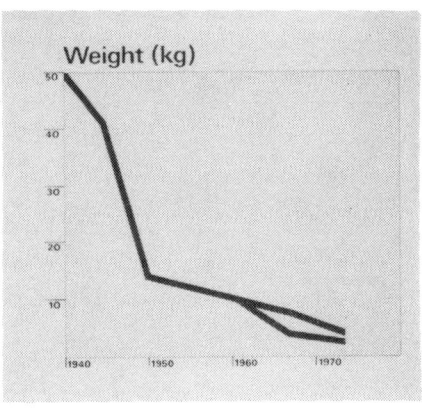

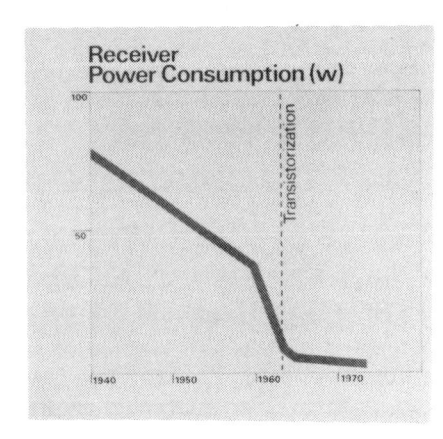

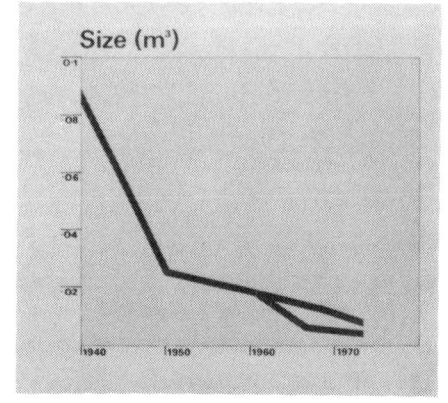

Figure 1.3

New component technology – new production processes

Most recently, we have seen the pace of technical progress dictated by:

- The use of large scale and very large scale integrated circuits; particularly transistor–transistor logic (TTL), emitter coupled logic (ECL), and low power complementary metal oxide semiconductor devices (CMOS).
- The use of surface mount production technology (the placing of components on, not through, the substrate).

Surface mount allows for higher component density and shorter interconnection paths, allowing for improved RF performance, together with cost-saving benefits – frequency coverage from 800 MHz used to require expensive (e.g. teflon) printed circuit boards, it is now possible to use standard fibreglass boards. Shorter connection paths help reduce electromagnetic interference and provide the basis for additional miniaturization, lower power consumption and weight reduction; circuit board areas are decreased by between 40 and 60 per cent. By reducing the number of solder joints, surface mount should also improve reliability (fewer dry joints).

New technology does not, however, guarantee a reduction in maintenance or service requirement.

Surface mount technology – implications

Surface mount, for example, depends on sophisticated adhesive techniques, chemical processes and soldering methods – wave, vapour phase and infra red – long term reliability of connections in practical field use still remains to be fully proven; long term reliability also depends on specialist circuit board design techniques dependent on stress analysis, specialist production techniques dependent on a clean production environment and critical temperature control techniques. Components are no longer isolated by their lead outs from the soldering environment. The soldered joint is the only connection, both electrical and mechanical.

Integrated circuit technology – implications

Integrated circuits are not immune to failure; intermetallic corrosion (the purple plague) can change operational parameters, such as noise performance and band width, linear chips can be particularly susceptible.

Different families of integrated circuits – CMOS, TTL, High Speed CMOS, ECL and Gallium Arsenide – are unmixable at an integrated level and require translator chips or interface components which are generally discrete. Hence power transistors, capacitors and diodes are still to be found in modern circuits as support components and are all susceptible to failure or degradation.

Impact on workshop repair procedures

New technology can also require changes in workshop repair procedure, for example:
Heat stress, replacing to component level with a soldering iron, particularly an iron without adequate thermostatic control, can cause heat stress to surrounding components, causing premature failure.
Re-work on surface mount requires specialist soldering irons (a jig in the shape of the component foot print, or a hot gas jet). Diagnosis requires specialist probes, component removal requires adhesive bonding to be softened.

Static, field effect transistors can fail or become noisy through handling, CMOS chips can become corrupted, degradation is not always immediately apparent. Static can also affect bipolar devices and passive components such as thick film transistor arrays.

Dust in a workshop can often be metallic and cause shorting or earth paths on PCBs.

Circuit design, modern design techniques, including CAD, use the inherent qualities of a circuit, inductance and capacitance, to the maximum. The effect of a solder blob needs to be trimmed out by a variable component; there may be insufficient adjustment available.

Software driven circuitry faults when they occur can be hard to simulate and create a problem of hybrid fault finding.

Demands on the technology

In addition, as technology improves, the demands made on the technology increase in parallel. As channel spacing reduces, RF parameters become increasingly critical; intermodulation problems are an example. Maintaining the operational parameters – modulation depth, frequency, power output, signal to noise ratio – through the operational life of the transceiver becomes increasingly critical.

New technology across the frequency spectrum

We tend to assume that most of the technical innovation is occurring in VHF/UHF, but it is worth remembering that new technology is opening up new applications in other areas of the spectrum, for example:

Long wave VLF/LF (3 kHz–300 kHz) and medium wave (300 kHz–3000 kHz)

- The vehicle location systems currently being introduced combine a VHF data link with long and medium wave synchronized navigation/location transmitters.

Short wave HF (3 MHz–30 MHz)

- New computer based propagation prediction methods potentially offer improved reception consistency with multi-path transmission/retransmission of messages.

New modulation techniques

- AM (amplitude modulation) has been decreasing in favour over the past years due to the relatively high power requirement on transmit, and the lack of discrimination between stronger and weaker signals.
- Spectrum efficient and power efficient AM modulation techniques, such as SSB (single side band) and SSBSC (single side band suppressed carrier), have been well proven in short wave applications, but to date have been unexploited in VHF, UHF.
- Component technology would now allow the realization of SSBSC at VHF without a major cost penalty (the necessary increase in stability of the IF oscillator can now be achieved relatively simply), but commercial constraints such as system incompatibility and manufacturing investment commitment to FM and conventional AM radios would need to be overcome.
- 'New' mobile radios using standard AM modulation are still being introduced – synthesized equipment with wide range switching capability (over 15 MHz) handling all standard signalling protocols – the water boards are UK users. AM will always have advantages for certain applications such as aviation where a weaker signal can be as, or more, important than a stronger signal within a given frequency band.
- FM will (presumably) continue to be popular, with the 'capture effect' (rejection of the weaker signal) generally proving to be an advantage in commercial usage, giving better SINAD (ratio of measured signal to noise and distortion) performance and resistance to static interference in most applications.

Data over radio

FFSK (fast frequency shift keying) is an increasingly commonplace method of data transmission (FFSK based on the conversion of binary 0s and 1s into a sub-carrier signal of 1200 Hz (1) and 1800 Hz (0). This does, however, produce additional problems not evident in a voice-transmitted system, as briefly outlined below.

Data transmission using FFSK requires radio (RF) parameters to be maintained at a higher standard, for example:

- *Harmonics* As with five tone systems, signals need to be of equal amplitude and free of harmonic distortion. Spurious sidebands can extend the radiated RF outside the licensing limits.
- *Frequency* It is difficult to keep a mobile receiver precisely on the exact frequency of a transmitter, giving rise to differential frequency error which can cause high bit error rates.

- *Modulation* The MPT* specification defines deviation, modulation depth and adjacent channel power parameters but a typical modulation meter would be calibrated for sine wave signals and would itself need calibration using a spectrum analyser to give a true indication of FFSK deviation or modulation depth.

Adding FFSK to an existing speech based radio system can create conflict for example:

- On FM, a PMR (private mobile radio) transmitter will commonly have limiting and filtering circuits to prevent frequency over deviation. The limiting and filtering circuits can cause interference to the FFSK signal.
- On AM, limiting and filtering to prevent modulation depth exceeding 100 per cent can cause FFSK interference.
- Bit timing stability can be a problem, particularly within radio networks where propagation throughout the system is a function of frequency (i.e. group delay). 'Intersymbol' distortion causes the displacement of the sub-carrier frequency components relative to each other.
- With 'derived' (as opposed to 'direct') land line links, frequency translation can produce an error of up to 6 Hz, which needs to be taken into account.
- Within any system, digital transmissions can be badly affected by Rayleigh fading and/or ignition interference, and/or doppler shift.

FFSK in quasi-synchronous systems

Quasi-synchronous systems, in which two or more transmitters radiate the same intelligence on the same channel, depend on the stability of carrier frequency off-sets, modulation level, delay differential and group delay distortion.

The sub-carrier data signals transmitted in a quasi-synchronous system, must have carefully controlled phase relationships, sub-carrier frequency errors are particularly disruptive.

FFSK in diversity reception

Networks with diversity reception (more than one receiving site within an area, with a system to select the strongest signal) can cause problems if the voting/switching system has not been designed to accommodate FFSK (speed of switching or switching during message transmission).

Digital speech systems

An analogue speech signal can be represented digitally, transmitted and reconstructed.

* See sources of help and advice (page 277).

- Processing analogue signals digitally, reduces the voltage, temperature and noise problems inherent in the analogue signal environment. At high frequencies it becomes increasingly difficult to design inductors and capacitors with sufficiently close component tolerances to maintain the accuracy and precision of the analogue signal.
- Digital processing requires no such precision components. Methods of digitization include the following:

Pulse code modulation

In practice it has been found that an accurate representation of an analogue wave form can be made if the signal is sampled at twice its maximum frequency; for example, in telephony with a frequency range of 300–3400 Hz, the sampling rate will need to be 8 kHz. Pulse coded modulation is used successfully in voice messaging systems, together with the variant of the technique known as continuously variable slope delta modulation, in which only the difference between one sample and the next is transmitted; both transmitter and receiver must retain a memory of the previous sample to reconstruct the signal.

Speech coders

More sophisticated speech coders have been developed for the PSTN digital network, including SBC (sub-band coders), and MELP (multiple excited linear predictive coders). The problem in all speech coders is to reproduce an adequate dynamic range. For instance, the analogue PSTN produces speech levels with a dynamic range of 40 dB – speech coders are currently limited to 25 dB. Problems common to line telephony and radio telephony include propagation delay and audible echoes. Problems specific to RF include the additional channel coding needed to protect the bit speech stream from fading on the radio path, and problems of co-channel interference from other users transmitting speech and data together.

Impact of innovation

In every area of mobile radio we can see technological changes occurring which are having a fundamental impact on the way in which radios are managed and maintained.

The technological impact can be expressed in a simple circle in which the knock-on effect of changing component technology can be shown on assembly techniques and circuit design, and in turn on radio operational problems – symptoms and cures.

Much of the technical focus today is on digital signal processing, in many

Mobile Radio Servicing Handbook

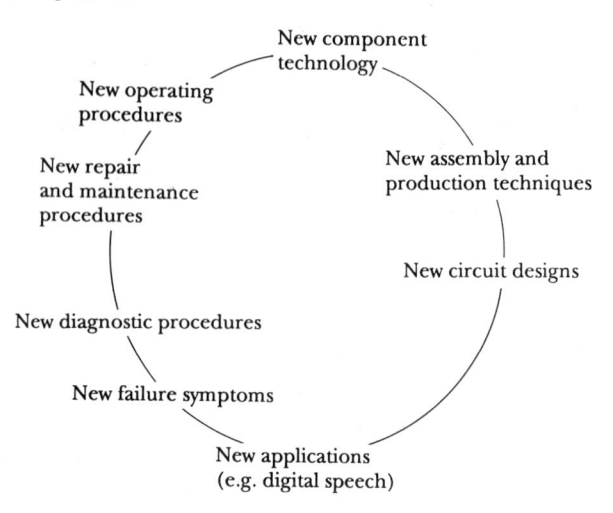

Figure 1.4 *The circle of causal relationships*

ways, the 'digital era' is being erroneously heralded as the 'cure for all evils', an attitude which tends to ignore the continuing importance of the radio link itself. We indulge ourselves in the next chapter by taking a step back and reviewing some of the basic fundamental principles of radio propagation, principles upon which our radio communication systems will, of necessity, continue to be based for the foreseeable future.

2 Propagation and frequency utilization

Frequency and wavelength

Remember that the sound output from humans is analogue. Speech using the sound of the human voice is created within the larynx by the vocal chords – a structure of bone, tissue and gristle which vibrates across a column of air expelled from the lungs to produce a wave of varying air pressure. These pressure waves are transmitted obviously via the mouth, but less obviously from the chest and upper part of the body, and are received by the ear, a membrane vibrating in sympathy with the pressure variation which converts the signal to a form which can be understood and interpreted by the brain.

We can collect the air pressure variations in a microphone – a moving diaphragm similar in principle to the human ear. Variations of *pitch* and *loudness* can be detected and converted into an electrical signal; a varying voltage which is 'modulated' onto the radio wave, or transmitted along a line before being restored to an acoustic signal by means of a loudspeaker or earphone.

It is important to establish the relationship between frequency and wavelength. Imagine yourself on a pier at the sea side. You count the waves moving past, *frequency*, and you measure *wavelength* as the distance between the crests of each successive wave.

In sound and radio waves, frequency is expressed in cycles per second – 10 Hz is equal to 10 cycles per second. For example, our hearing frequency range extends from 15 Hz to 15 kHz (15 000 Hz). An intelligent audible voice signal can be conveyed within a restricted range from 300 Hz to 3k Hz (3 000 Hz).

In any wave propagation, wavelength is equal to:

$$\frac{\text{Velocity of transmission}}{\text{Frequency}}$$

For example, a sound signal has a velocity of 760 mph (1216 kph), or 1114 ft/sec at ground level, and a 50 Hz signal therefore has a wavelength of

$$\frac{1114}{50} = 22.7 \text{ ft}$$

A radio signal is an electro-magnetic radiation and has a velocity of 3×10^8 metres/sec (186,000 miles/sec).

Hence a 300 KHz signal has a wavelength of

$$\frac{3 \times 10^8}{3 \times 10^5} = 10^3 \text{ metres}$$

Sound waves are easily absorbed (attenuated) and can only travel limited distances. Radio waves can be made to travel much further – the basis for radio transmission. Radio waves also travel much faster. Light takes 8.5 minutes to reach us from the sun (93 million miles). In 8.5 minutes, a sound wave would have travelled 100 miles.

By transferring sound waves to radio wave frequencies, we have the medium for fast, long distance speech communication. The 'signal' is modulated on the radio wave at a specific frequency, taking up space on one or either side – the *'frequency bandwith'*.

Electro-magnetic radiation covers a very wide frequency range. The range suitable for radio transmission at the present state of the art extends approximately from 30 kHz to 30 GHz. Higher frequencies embrace the radiation of heat and light.

The radio spectrum can be divided according to the range and nomenclature shown below:

		Frequency	Wavelength
Long wave	VLF/LF	3 kHz–300 kHz	(100 km–1 km)
Medium wave	MF	300 kHz–3000 kHz (3 MHz)	(1000 m–100 m)
Short wave	HF	3 MHz–30 MHz	(100 m–10m)
Very high frequency	VHF	30 MHz–300 MHz	(10 m–1m)
Ultra high frequency	UHF	300 MHz–3 GHz	(1000 mm–100 mm)
Micro wave frequencies	SHF	3 GHz–30 GHz	(100 mm–10mm)

Standing waves

Wave propagation has two constitutents – potential (electric field) and current (magnetic field) for an EM propagation, and incremental velocity and pressure for an acoustic wave. In a uniform media these are propagated at the same velocity, attenuated equally, remain in phase and have a constant ratio equal to the characteristic impedance of the transmitting media. The level of transmitted power is proportional to the product of the two constituents.

If the transmission meets a discontinuity or obstruction, representing a change in characteristic impedance of the media, reflection takes place. This could be a building or a hill for free space radiation or an open ended or short circuited transmission line (EM wave), or a closed or open end to a pipe (acoustic wave). Since the reflected wave traces its original path, but in the

Propagation and frequency utilization

opposite direction, the phase relationship of incident and reflected wave vary along the path, giving rise to progressive change in amplitude, with maximum or minimum (referred to as antinodes and nodes) occurring at quarter wavelength intervals. Such wave patterns are known as 'standing waves'.

An antinode voltage always coincides with a node of current and vice versa, hence a quarter or half wave line, with a short circuit or open circuit at the far end, will act as a frequency selective resonator at its input end.

This phenomenon is of great importance both for acoustic and EM propagation. The pipe resonator with standing waves set up in the pipes is the basis for nearly all wind instruments, including, of course, the organ. An organ pipe half a wavelength long with an open end will present a low impedance at the reed at its resonant frequency, ie a 50 Hz note, having a wavelength of 22.7 feet, requires an open ended pipe of 11.35 feet. The same phenomenon is the basis of antenna design.

For instance, a quarter wavelength dipole must have a voltage antinode and current node at its open end, which is brought about by a reflected wave of current equal to, but opposite in phase to, the incident wave, and a reflected wave of voltage equal to and in phase with the incident wave. When these reflected waves reach the input end of the dipole, the current wave will add to the incident wave and the voltage wave will cancel, giving rise to a voltage node and current antinode at the resonant frequency. For a loss free antenna, the input impedance at resonance would be zero, but losses are caused by loss in the antenna and the radiated energy, expressed in terms of 'radiation resistance'. This gives a finite input resistance of resonance, which is the resistance to which the transmitter should be matched. It should be appreciated that when the antenna is matched in this way, the terminating impedance at the send end is much lower than the characteristic impedance of the line along which the standing is generated. Progressive reflections then occur from each end, so that the voltage antinodes build up to a value very much higher than the doubling that occurred at the first reflection.

By the same reasoning a half wave line will have the opposite effect, producing a high impedance anti-resonance at the input end with an open circuit far end, as will a twin quarter wavelength line or coaxial cable with its far end short circuited. The quarter wavelength line in fact inverts the terminating impedance, an open circuit at the far end appearing as a short circuit at the input end. It can be shown that the input impedance =

$$\frac{(\text{characteristic impedance})^2}{\text{terminating impedance}}$$

and this principle is used to provide impedance matching.

Mobile Radio Servicing Handbook
Spectrum utilization

Channel width and channel availability

- With over 80 per cent of the frequency spectrum allocated to non-commercial (military or government) applications, the pressure on the remainder of the available spectrum is intense and has resulted in research into use of parts of the spectrum other than VHF/UHF, for instance long term research into mobile cellular systems based on frequencies up to 60 GHz.
- Given that channel availability increases with frequency, it is more likely that long term solutions will be found at higher, rather than lower, frequencies.
- With speech, the minimum bandwith required is equal to twice the highest speech frequency (generally 2700 Hz) to which we need to add 6000 Hz of separation bandwith between adjacent radio channels, plus a guard band to allow for frequency drift or limitations of receiver sensitivity, giving us our normal radio frequency speech channels of 25 or 12.5 kHz.

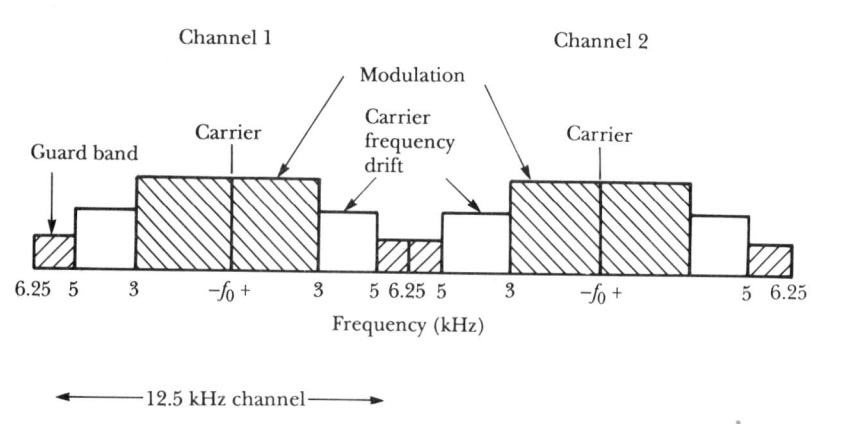

Figure 2.1

Taking a 12.5 kHz channel as an example, we can see that channel availability increases with frequency.

On MF medium wave (300 kHz–3000 kHz giving 2700 available kilohertz), the number of channels available will equal

$$\frac{2700 \text{ kHz}}{12.5 \text{ kHz}} = 216$$

On VHF (30 000 kHz–300 000 kHz giving 270 000 available kilohertz), the number of channels available will equal

$$\frac{270\,000 \text{ kHz}}{12.5 \text{ kHz}} = 21\,600$$

The short distance limitations of VHF and UHF are, of course, well suited to two-way radio in the sense that frequencies can be re-used within reasonably close proximity – the shortcoming becomes an advantage.

Transmission to optical wavelengths

The principle of channel availability increasing with frequency remains valid at optical wavelengths (400–800 nanometers), hence the high channel carrying capacity of optical fibre. Transmission of information using light wave in free space would be attractive were it not for the high attenuation from dust, sand, snow and fog, and line of sight limitations.

Spectrum propagation characteristics

The propagation characteristics have a profound effect on the choice of suitable frequency band for a particular purpose. Propagation takes place between the earth's surface and the ionised layer in the upper stratosphere surrounding the earth, both of which have varying effects dependent on frequency.

The earth (including water) is a reasonably good conductor and the currents set up by an EM transmission dissipate the transmitted energy, particularly if the transmission is horizontally polarized (electric field parallel to the earth). Absorption of energy increases with frequency, with the result that VLF/LF can cover large distances (several thousand miles in suitable circumstances) with the ground wave. Also, due to the low earth losses, the transmission will penetrate earth or sea for a short distance, and is the only band capable of providing communication to such situations. The range is also assisted by the ability of the transmission to follow the earth's curvature.

As frequency increases, so the rate of energy absorption of the ground wave increases, resulting in progressively decreasing range for the ground wave. Depending on ground conditions, the absorption reduces effective range progressively to hundreds of miles at MF, to tens of miles at VHF and, effectively, line of sight at UHF.

The effect of the ionised layers is more complex. The density of free electrons in the upper atmosphere is controlled by sunlight, and therefore increases considerably in daylight and during the summer months, giving rise to both diurnal and seasonal variations in transmissions.

Free space normally acts as a dielectric, but the presence of free electrons reduces the dielectric constant by a factor

$$(1 - \frac{\omega_c^2}{\omega^2})$$

where ω_c is known as the 'critical frequency', and is directly proportional to the concentration of free electrons.

For values of ω below ω_c, the dielectric constant becomes negative, and the media ceases to act as a dielectric, becoming effectively a conductor and absorbing the energy of any EM wave entering the layer. In terms of frequency, this occurs at VLF/LF at night, and VLF/LF and MF during the day (when ionisation is higher and ω_c therefore higher). At frequencies where the ground wave has the highest range, therefore, the sky wave is absorbed, resulting in reliable fade free reception over relatively large distances. For frequencies above ω_c the dielectric constant increases with frequency, becoming equal to that of free space for $\omega >> \omega_c$ (ie SHF night and day, and UHF at night). At these high frequencies, therefore, the ionised layer has negligible effect, and radiation penetrates the layer unaffected. This makes these bands suitable for satellite communication.

At intermediate frequencies, the radiated wave passes from the un-ionised atmosphere into the ionised layer of lower dielectric constant. The result is similar to that of light entering water, the velocity increasing and the wavefront bending. The angle of the transmission to the vertical for which the wavefront is best horizontal, is known as the critical angle (ϕ_{crit}), and for angles greater than this, reflection takes place. It can be shown that:

$$\cos \phi_{crit} = \frac{\omega_c}{\omega}$$

The significance of this is that, at frequencies above ωc, the ionised layer will reflect the transmission, but the angle of delineation from the vertical must be greater than:

$$\cos^{-1} \frac{\omega_c}{\omega}$$

For instance, at a frequence $2\omega_c$, ϕ_{crit} is 60°, and that part of the transmission with an elevation of less than 30° will be reflected. For a frequency $5\omega_c$, ϕ_{crit} is 78°, and only signals with an elevation of less than 12° would be reflected.

The lower the angle of elevation of the reflected wave, the greater will be the distance from the transmitter before it reaches the earth's surface again. In the HF band, where the reflected wave is of greatest significance, the ground wave attenuation increases with increasing frequency, and the point at which the reflected wave appears becomes more distant with increasing frequency. This can give rise to a 'dead' area between the two signals, referred to as the 'skip distance'. As frequency increases with the VHF band, the ground wave attenuation becomes more rapid, and the angle of elevation of the reflected wave is so low that it never returns to earth, due to attenuation and the earth's curvature.

It should also be born in mind that, at frequencies near to ω_c where reflection is occurring at steep angles of elevation, multiple reflection between the ionised layer and the earth's surface can occur. The ionised layer is extremely mobile, changing very quickly. The reflected signals can therefore vary rapidly and the sidebands can vary relative to the carrier, giving rise to selective fading and very distorted speech.

Frequency allocation

We need to look at the various factors influencing the allocation of frequencies.

VLF/LF can be transmitted very long distances due mainly to their very low attenuation in physical mediums, e.g. air, earth, water and their ability to follow the curvature of the earth. Disadvantages are that antennas (whose size is directly related to wavelength) are very large. Also not so many channels are available. Consider a channel width of 5 kHz – a narrow chanel – at 200 kHz this channel represents some 2.5 per cent of the frequency in use!

MF and above cannot penetrate earth and water for any usable distance and so can be considered as wavelengths for atmospheric use. Propagation is mainly by reflection from the earth's ionised layers and ground effects, i.e. direct path. These were historically some of the first wavelengths to be developed for commercial use and remain so today. The antenna systems required are medium to large and band utilisation is not efficient if AM is used for speech transmission. Signals will typically travel a few hundred miles in the daytime, several thousand miles at night.

HF was first demonstrated as a practical long distance communication band by radio amateurs in the 1920s. Power required is relatively low (less than 150 watts) and antenna systems can be quite compact. Transmission mode is by reflection from the ionosphere, and many developments have given rise to efficient techniques such as double and single side band.

VHF/UHF Short distance 'line-of-sight' working (but can, under certain atmospheric conditions, go long distances though not sufficiently frequently or predictably). Low cost hardware and semi-miniature antennas, hence compact equipment. Sufficient channel space is available for local and 'networked' communications.

SHF is highly directional, needing sophisticated antennas and expensive electronics. Forms reliable 'line-of-sight', difficult to interfere with communication. Carrier frequency to band width ratio is high enough for even the fastest data communication.

Required use and application for VHF/UHF

Radio waves at VHF/UHF and SHF behave very similarly to light waves. They travel in straight lines – objects which cast a shadow to light cast a shadow to radio waves – similarly reflection, refraction and diffraction allow radio signals to travel over the visual horizon and into areas of shadow.

The following general rules apply:

1 As the frequency increases, range decreases but so does the ambient electrical noise.
2 Reflections from buildings, particularly ferro-concrete, increase with frequency.

The result is that:

1 Low VHF bands (up to 88 MHz) are generally most suitable for large rural areas.
2 High VHF bands (148–225 MHz) are generally most suitable for mixed urban and rural areas.
3 UHF bands are generally most suitable for urban work.

Compromises are, however, often necessary. Cellular, for example, at 900+ MHz, was chosen for good urban coverage and channel availability.

Transmission at millimetre wavelengths 30–300 GHz

At frequencies above 30 GHz, propagation is severely limited by attenuation. It is generally considered that conventional cellular technologies might be applicable up to 4 GHz. Micro-cellular technologies at 40–60 GHz, would require a transmitter on every lamp post – radiating the street both with light and radio waves.

Attenuation at these frequencies consists of:

- *Resonant absorption* Molecules of oxygen and water vapour absorb radio waves at particular frequencies, with an attenuation peak for water vapour at 22 and 183 GHz and an attenuation peak for oxygen at 60 and 119 GHz. The 60 GHz oxygen attenuation peak for example produces attenuation of 15 dB per kilometre. Practical transmission length, and hence potential frequency re-use, is little more than 100 metres. The parts of the spectrum where oxygen and water attenuation are low are known as 'atmospheric windows' and occur at 37, 97, 137 and 210 GHz

- *Non-resonant absorption* Signals are scattered by raindrops, hail and snow. Being non-resonant, the effect increases with frequency, for example as the wavelength reduces towards the size of a raindrop. Rain, hail and snow can, therefore, have a significant effect on link quality.

Because of the small droplet sizes, fog is not a problem at millimetre wavelengths but creates severe attenuation at infra red and optical wavelengths.

Established applications using millimetre wavelengths already include electronic news gathering (one-way video) at 40 GHz, and a number of military uses – for example missile guidance.

Commercial mobile radio applications will depend on the development of low cost, high volume component technology capable of working at these frequencies – current development work is being focused on indium phosphide and gallium arsenide monolithic integrated circuits.

Transmission via satellite links

Satellites can be used in theory as an effective wide area coverage repeater, the limitation being cost and no frequency re-use. Ship to shore and aviation communications are provided by the International Maritime Satellite Organisation (INMARSAT) at 1.4 and 1.6 GHz. Satellites work within a band of frequencies from 100 MHz (30 cm) below which signals become absorbed by the ionosphere up to 30 GHz, above which the atmosphere creates high attenuation. The most popular frequencies are 1–18 GHz with satellite TV at 11 GHz.

For land mobile radio, geostationary satellites would provide good wide area coverage in equatorial and moderate latitudes, but (as with satellite TV), the low elevation angles in Northern Europe of 30° or so would result in poor, limited line of site reception.

Satellites in the Molniya (12 hour) or Tundra (24 hour) elliptical orbits provide higher elevation angles, but with the disadvantage that more than one satellite is needed to provide 24-hour coverage.

The 'footprint' of the elliptical orbit satellite would typically be 200–300 kilometres, placing significant limits on frequency re-use (compare 2–3 kilometres for cellular, or 200–300 metres for micro-cellular).

An alternative would be to use a multiple spot beam to provide localised and switchable coverage.

Research work* undertaken at the Rutherford Appleton Laboratory suggests that a 40 beam spot satellite could support a potential mobile radio user population of 2 million users in Europe with a digitised speech data rate of 4.8 kbps, 10 MHz of spectrum and 50,000 users per beam.

Sharing of satellites with other services (eg broadcasting on the existing Inmarsat network) may also be a practical and potentially cost effective option.

Legislation on frequency allocation

This, for practical purposes, is the absolute determining factor on our choice of frequencies. Allocation is decided on a first order basis at international meetings of the World Administrative Radio Conference (WARC), organised by the International Telecommunication Union, part of the United Nations with 160 member countries, and on a second order by national and local

* Research work by J. R. Norbury, Rutherford Appleton Laboratory Telecomms, September 1988.

politics. There is also an International Frequency Registration Board which oversees radio frequency spectrum on an international basis.

The 1979 WARC conference had a major effect on the mobile communications market in the 1980s. The Band I 405 line television frequencies were released for PMR (41–47 MHz, 47–48 MHz on a permitted basis). The decision was also made to release Band III 405 line frequencies (174–216 MHz, 223–230 MHz). The French TV, Belgian and Irish Broadcasting Authorities are still transmitting in these bands and may cause some interference in the South of England and Western coastal areas.

The African countries at the same time succeeded in re-allocating frequencies from 87.5–108 MHz for VHF/FM sound broadcasting, which has meant UK police and fire services (currently at 97.6 and 102.1 MHz) changing frequency by 1989.

In the same way, the UK PMR systems using 104–108 MHz will have to be re-allocated by 1995, a replacement opportunity for system/service providers.

On a local basis, legislation also has a fundamental impact. For example, it wasn't until 1979 that VHF links to a base station were permitted as an alternative to land lines, opening up the potential for lower cost community repeater systems. James Merriman's review of the UK radio spectrum in 1982 similarly influenced the government's decisions on national (as opposed to international) allocation between military, utility and commercial applications.

Commercially, the UK market continues to be dictated by DTI licensing; the two major landmarks have been the Wireless Telegraphy Act of 1949 for land mobile (PMR) applications, and the Telecommunications Act of 1984 (where companies are providing services to third parties).

Government policy can also fundamentlly influence the characteristics of each country's mobile radio industry. The establishment of competing service provider networks in the UK for cellular and Band III is largely credited as the reason for the fast growth of these services in the UK relative to our European neighbours, where, with the exception of Scandinavia, cellular technologies have been slow to be applied.

Radio frequency chart – spectrum allocation

The positioning of mobile radio within the radio spectrum is outlined in detail in the Radio Frequency Chart (Appendix III) and Private Mobile Radio Bands (Appendix V).

290 MHz of spectrum is currently available between 30 and 960 MHz on the basis of worldwide WARC allocations. Prior to the Merriman Report, only 51 MHz was allocated to mobile radio in the UK. UK mobile radio allocation is now just over 150 MHz, including cellular (30 MHz) and Band III, i.e. approximately fifteen per cent of the VHF/UHF spectrum.

3 System choices

Before dealing with the detail of receiver and transmitter fundamentals we need to review the system choices available to the user and clarify some of the system descriptions currently being used in the UK. To compare the cellular systems used in other parts of Europe see the specialist section on cellular radio (Chapter 14).

Cellular radio

This is the closest equivalent to the standard telephone service with direct dialling into the public sector telephone network. Connection is made via a radio link to a localized base station. Two national network operators (Cellnet and Vodafone) provide coverage. Operating at a relatively high frequency gives transmission clarity and data transmission capability.

Three basic types of handset are available to users. The standard car phone is permanently installed, extras include 'hands free' operation, normally a dashmounted speaker/microphone. A 'transportable' has the same features, but can also be used remotely from the vehicle (units have integral battery packs). A 'portable' is fully self-contained. Car phones and transportables can be switched between vehicles, provided the vehicles have been fitted with the required equipment – handset cradle, wiring, harness and antenna. The user also normally has the choice of transferring between the two networks.

Additional facilities available include diversion of incoming calls, 'call waiting', and conferencing.

Channel spacing 25 kHz
Modulation method FM

Infrastructure
600–700 base stations to provide 98% coverage

System 4

System 4 was the predecessor of cellular radio, with a London service first established in 1965. Run by British Telecom, recent capital investment has improved coverage and service standards. Geographic coverage is as extensive as the cellular system, and the user has the choice of national coverage or selecting a discrete zone – South East, Midlands, North or Scotland for example.

System 4 advantages over cellular are therefore, in certain circumstances, geographic (larger area coverage) or lower cost where only limited local zone coverage is required. In London only outgoing calls can be made. System 4 mobiles, when wishing to make a call, will poll all channels until one is found that is free, whereas in cellular, channels are dynamically allocated. System 4 connection times may, in certain circumstances, be longer.

	Infrastructure
Channel spacing 12.5 kHz	130 base stations nationwide
Modulation method FM	

Wide area paging

As with System 4, wide area paging coverage in the UK is divided into zones. Network operators/service providers give users a choice of local coverage (within a zone), wide area coverage, or (with certain systems) nationwide coverage, accessing the system by direct dialling, via a terminal, computer and/or operator.

The types of pagers available include:

1 Tone pagers providing an audible or visual alert to the user.
2 Voice pagers, a ten second message in the voice of the caller.
3 Number pagers, transmission of telephone numbers or a pre-arranged code.
4 Word pagers, a full alpha-numeric message.

Pagers can be linked into other services such as voice message, and/or cellular call diversion. Units are also available for use in hazardous environments where gas or chemicals may be present.

	Infrastructure
Channel spacing 25 or 12.5 kHz	British Telecom's National Network
Modulation method FM	400 Base Stations, 3 × 25 kHz channels supporting 450 000 users

On-site paging

This is paging within a defined area, site or building. Generally the lowest cost, most localised form of mobile communication, on-site paging uses low power handsets with a base station or induction loop integral to the building structure, including alpha-numeric paging for hospital/emergency applications.

Channel spacing   25 or 12.5 kHz
Modulation method   FM

Private mobile radio (PMR)

Most PMR applications are mobiles talking to base or base talking to mobile (e.g. salesman or serviceman to office) in a 'closed' (hence 'private') user group.

Private mobile radio is generally simplex or semi-duplex (one speaker at a time – PTT (press to talk)) in operation. There are four main types.

Single systems

A single base station giving local coverage (typically 5–10 mile radius) provides direct access to a number of mobiles. This type of system involves the user in a fixed capital investment cost, offset thereafter by lower running costs limited to an annual licence fee and routine maintenance.

Improved area coverage (up to 50 mile radius) can be obtained by locating the base station at a remote site (a hilltop for example) with connection by telephone land line or fixed radio link to the controller's premises.

The user has the responsibility of obtaining and complying with a licence issued by the DTI and administered on a local basis by the Radio Investigation Service.

Community systems

Basically a shared facility, administered by a third party, reducing the capital cost investment of a single system. The third party 'service provider' supplies hardware and air time to the user as required, generally an 'access fee' is charged on a monthly basis, or less commonly, air time is charged per call. Additional services can be available, including limited access to the PSTN (public telephone network), secretarial/message taking and group calling.

Trunked systems

The technique of trunking allows a number of base stations to be linked together typically by telephone land line or fixed radio links (and more recently optical fibre links). The linking of trunked base stations can provide increased coverage.

Adding computer control to a trunking network also increases the efficiency with which the allocated frequency channels are utilised, with queueing and priority allocation minimizing access/connection times. Some large national organisations have their own trunked networks providing wide area coverage, similarly community systems are now available in which groups of users can 'sign on' to a network, paying for access and/or air time and additional services as required, with a choice of *local* or *regional* coverage.

Band III

The government's recent release of the black and white TV transmission frequencies (Band III) has resulted in the formation of two separate competitive national consortia and a number of local consortia, set up to provide a choice of local, regional and nationwide trunked radio services. As with cellular, the national network operators do not deal directly with the public, but provide access and air time to retailers, who in turn supply the end user with equipment and service, including installation, connection, billing and maintenance.

Significant spectrum efficiency benefits are claimed by the Band III network operators (see trunking section Chapter 14) with the added capability of providing service in areas within an existing high density of PMR users, e.g. areas with more than thirty radio base stations per square kilometre.

Band III charging is either on a coverage basis (similar to the concept of paging zones and areas), the user opting for local coverage, wider area coverage (regional), or national coverage, or on the basis of air time.

Channel spacing   25 and 12.5 kHz   (Band III – 12.5 kHz FM)
Modulation method   FM

CT2 'phone zone' or 'telepoint'

The new de facto standard being adopted in the UK and (now) other parts of Europe for the next generation of cordless telephones.

It provides a mobile radio handset which will work as existing cordless phones, but can also be used to make outgoing calls through localized base stations, e.g. at railway stations.

32 KBPS digitized speech

Pan-European (Group Special Mobile) digital cellular system

A new fully digital speech and data cellular network to be implemented across Europe from 1991 onwards – will allow mobiles to roam from country to country. The Pan-European GSM network will be introduced alongside the existing FM analogue cellular systems.

Channel spacing – 200 KHz
Digital voice coding at 13 kbps
Modulation method – gaussian minimum shift keying

Service and repair requirements for each system

As we shall see from the following chapters, most of the diagnostic principles and repair procedures are applicable to any radio system using traditional AM or FM modulation techniques. *Note:* The RF analogue networks being installed today – cellular radio and the trunking networks included – have a life expectancy of at least 25–30 years.

The digital radio systems – CT2 and the Pan-European GSM cellular system (fully described in Chapter 15) – still depend on the integrity and quality of the RF link.

A detailed understanding of the principles and practice of RF communication, as set out in the following pages, remains an essential requirement for the mobile radio engineer.

4 Principles of RF communication

Introduction

A basic RF communication system consists of a transmitter coupled to an antenna, a receiver coupled to an antenna, and a suitable propagation condition for the signal passing between the two.

The function of the transmitter is to convert the information to be communicated into an RF signal suitable for application to the antenna. Given sufficient power and suitable conditions, the radiated signal will cover the area and distance required of the system.

The function of the receiver is to process the signal intercepted by the antenna, validate it and convert it back to the original form presented to the transmitter.

Concept of modulation

It was seen in Chapter 2 that speech consists of waves of varying air pressure. As this effect is rapidly attenuated with distance, speech communication is impractical at more than a few feet, and certainly is not private. It is clearly unsuitable for PMR communication purposes.

This varying air pressure can be converted into an electrical energy or signal in a microphone. The signal can be converted back into waves of varying air pressure – our original speech – by means of headphones or a loud speaker. The microphone and loud speaker can be connected together by wire and, as long as any losses are made good, en-route communication is possible at any distance. However, this system does not fulfill the criteria of 'mobile'.

To be truly private and mobile, radio waves are used as a medium. It is possible in the simplest case to convert the electrical speech signal into a radio wave of the same frequency (300–3400 Hz). The disadvantages of this are that all communication would be on the same frequency – chaos, and the antenna would be gigantic. To make a compact system with sufficient separate (i.e. non-interacting) communication possibilities or channels we need to look at using a higher radio frequency. From the earlier considerations of propagation characteristics it was seen that PMR communication is optimum at VHF and UHF.

Principles of RF communication

To summarize, speech is in the form of an electrical signal at a frequency of 300 Hz to 3400 Hz and the radio signal at a frequency of 30 MHz to 1000 MHz. The speech signal must be superimposed onto the radio signal and the combined wave used to carry the information to the distant receiver. This combining of the two signals is achieved by a process known as modulation. Conversely separating the two signals to yield the original speech signal is called demodulation.

Modulation is the process by which intelligence is added to an RF carrier.

Aims of modulation

1. Transfer of information with no distortion.
2. Efficiency of conversion. Modulation must be achieved with the minimum loss of power.
3. Efficient use of RF spectrum. It has been seen that the amount of RF bandwidth available is strictly rationed. For this reason a carrier with as narrow a bandwidth as possible is used.
4. Minimize the RF power requirement – to keep the equipment simple, the heat it generates to a minimum and to ensure a long battery life, all the RF power should be used in carrying the information.

Types of modulation

Types of modulation in common use include:

Amplitude modulation	AM
Double side band modulation	DSB
Single side band modulation	SSB
Frequency modulation	FM
Phase modulation	PM
Pulse coded modulation	PCM

Amplitude modulation

In this modulation technique the amplitude (or strength) of the carrier wave is varied in sympathy with the audio wave.

Mobile Radio Servicing Handbook

Input or audio

Unmodulated carrier

Modulated carrier

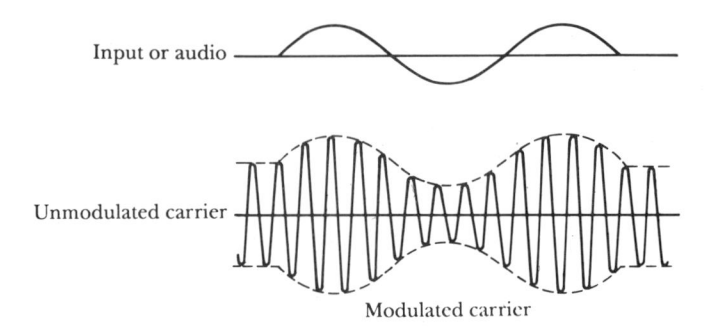

Figure 4.1

Figure 4.1 shows a sinusoidual carrier wave being modulated by a single low frequency sine wave. In practice the low frequency waveform would consist of a number of sine waves 'mixed' together to create a speech sound and would last for the duration of each spoken word. It can be seen that the amount by which the carrier level is increased or decreased in strength is directly related to the amplitude of the modulating waveform. The measurement of the modulation amount (or depth) is given as a percentage ratio of the maximum amplitude to the minimum amplitude.

$$m\% = \frac{\text{maximum amplitude} - \text{minimum amplitude}}{\text{maximum amplitude} + \text{minimum amplitude}} \times 100\%$$

If the modulation wave is increased sufficiently in amplitude, the troughs of the carrier will be reduced to zero. At this point 100 per cent modulation occurs, an undesirable state for the communication system as will be seen later.

This process of symmetrical amplitude modulation shown in Figure 4.1 produces a band of frequencies above the unmodulated carrier frequency and a band of frequencies below the unmodulated carrier frequency. The extent of these sidebands is directly a function of the modulating frequencies, i.e. they extend as far above and below the carrier frequency as the highest frequencies occurring in the modulating waveform. Hence, to use the radio spectrum as efficiently as possible it is necessary to contain the speech frequencies in the minimum possible bandwidth. Figure 4.2 shows the carrier, the upper sideband and the lower sideband resulting from the modulation process.

In AM it is this composite waveform that is to be transmitted. All three parts require power to create the carrier waveform. Under maximum modulation operation, two thirds of the total power is contained in the carrier (fc) alone and only one third in the sidebands. As the carrier contains no information, this obviously is wasteful of power. By means of circuitry in the transmitter it is possible to remove the carrier so that only the sidebands containing the

Principles of RF communication

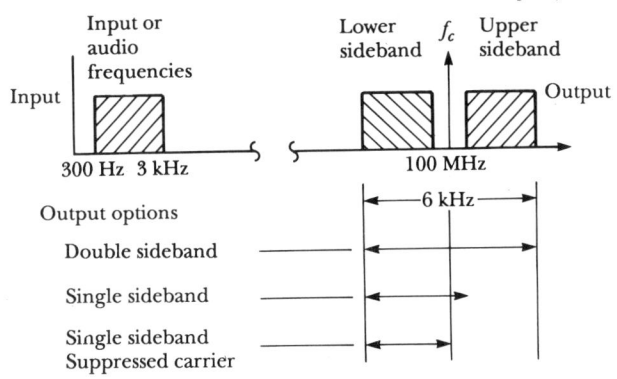

Figure 4.2 *Amplitude modulation – sidebands*

information require power for transmission. This more efficient system is called double sideband suppressed carrier, commonly abbreviated to DSBSC. Further examination of Figure 4.2 shows that both the upper and lower sidebands contain the same information. Again, by means of suppression in the transmitter it is possible to remove either sideband or, in the limit, either sideband and the carrier. The former is single sideband transmission (SSB), and the latter single sideband suppressed carrier (SSBSC).

The removal of one sideband and retention of the carrier is not a satisfactory method as, although it saves bandwidth, it still wastes transmitter power by radiating the carrier, and also results in distorted demodulation if a simple detector is used. SSBSC saves bandwidth and power, and in contrast to DSBSC, the carrier need not be introduced in exact phase, and in fact a small frequency error can be tolerated.

A further advantage of suppressed carrier systems is increased insensitivity to external interference. The disadvantages of these systems is the increased circuit complexity and hence cost of both receiver and transmitter. These systems are found primarily in the HF bands on point-to-point communications but suppressed carrier and sideband techniques are constantly under review for PMR systems.

Frequency modulation (FM)

FM is the prime modulation technique used in mobile communications. Frequency modulation varies the frequency of a carrier wave in proportion to a modulated signal. The carrier amplitude of an FM wave is kept constant during modulation, and so the power associated with an FM wave is constant. During modulation, the carrier frequency increases or decreases when the modulation voltage becomes positive and changes in the opposite direction when the modulation voltage becomes negative.

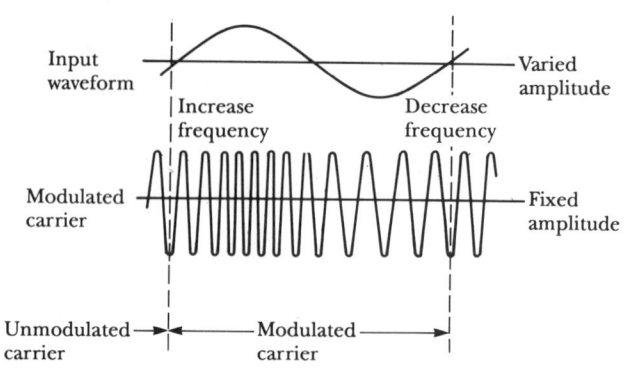

Figure 4.3 *Frequency modulation (FM)*

The deviation of frequency from its unmodulated value is known as the 'deviation' (ω_d), and the frequency deviation for the equivalent of 100 percent modulation is known as the 'peak deviation'. It should be noted that the deviation is proportional to the *amplitude* and not the *frequency* of the modulation frequency (ω_m).

ω_m = Modulation Frequency = 2π fm   Both measured in
ω_d = Deviation = 2π fd   radians per second.

The deviation divided by the modulation frequency is known as the 'modulation index' (m).

As with AM, the modulated waveform can be analysed into a number of constant amplitude signals, which determine the bandwidth requirement. With FM, these signals comprise the carrier and a series of sidebands spaced at the modulation frequency, and decreasing in amplitude. The number of these sidebands required is proportional to the modulation index, and a satisfactory result is obtained if m + 1 sidebands are transmitted. Since these sidebands are required on both sides of the carrier, and are spaced at the modulation frequency, this gives a bandwidth requirement of:

$$2(\omega_d + \omega_m).$$

The choice of modulation index together with circuits which determine whether a large or small number of sidebands are allowed through the system, determine the highest and lowest modulating frequencies which the system can process. A large number of sidebands allowing a wide range of modulation frequencies, typically tens of kHz, is referred to as a broad band system. Conversely for the speech band used in PMR communications a narrow band system is used.

Wide band signals are used for broadcast quality transmission where it is required to transfer a large number of frequencies for music, for instance from the low frequency organ pipe up to high frequency violin notes. The small

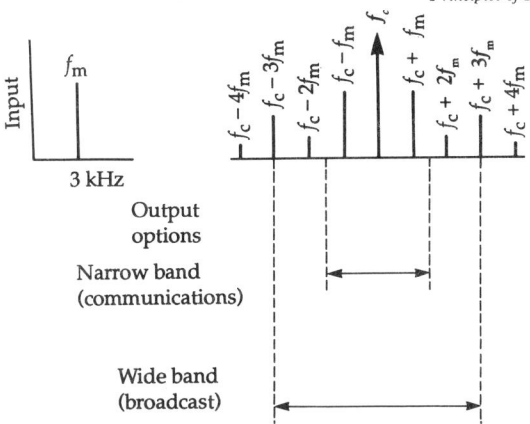

Figure 4.4 *Frequency modulation sidebands*

modulation index will produce a narrow band system, and it is this that is used for PMR communication.

As the modulating waveform can be made to have a larger and larger effect on the carrier, the maximum modulation depth is determined by system capability. This means that it is possible to increase the modulation index to a point beyond the capability of the transmitter and receiver circuits to handle the resulting carrier without undue distortion. This situation is extremely undesirable and will result in interference to adjacent communication channels and distortion of received signals.

The speech band has been defined as covering 300 Hz to 3400 Hz. The energy distribution across this band when considering a normal composite speech waveform is not even, but falls towards the upper end of the band, whereas in an FM system as described, the noise appearing at the output of the FM demodulator or detector increases with audio frequency. This means that the signal-to-noise ratio (a measure of signal intelligibility or clarity) worsens at the upper end of the speech band. It is possible however to utilise 'a trick' to overcome this problem. In the speech frequency sections of the transmitter a circuit can be included which will increase the amplitudes of the higher speech frequencies with respect to the lower frequencies, so that the modulation index to noise ratio improves as the modulating frequency increases. This process is known as *pre-emphasis*. At the receiver, in order to restore the sound to its correct balance, there is a circuit with the inverse function. This circuit will attenuate the higher frequencies with respect to the lower frequencies. This process is known as *de-emphasis*. It is important that the degree of pre-emphasis and de-emphasis is matched in a system to ensure no distortion is introduced. A standard characteristic is used by PMR and is typically 6 dB per octave. There is also a CCIR recommended characteristic for multi-channel radio relay systems where the degree of pre-emphasis varies across the channels.

Advantages and disadvantages of AM and FM

The disadvantages of AM include its susceptibility to man-made electrical interference or noise, its inefficient use of power and its relatively limited dynamic range of modulation compared with FM. The first two points can be improved by the use of DSBSC techniques, but at the expense of more costly circuit implementation.

The prime advantage of AM is the simple modulators and detectors required to use this mode. It also requires a smaller bandwidth. $2\omega_m$ for DSB or ω_m for SSBSC, against $2(\omega_m + \omega_d)$ for FM.

FM has several advantages over AM techniques. FM is far less affected by man-made interference as these effects tend to alter only the amplitude of the signal and can be suppressed by limiting before demodulation. Also an FM receiver exhibits a capture effect. This is the ability to resolve only the stronger of two signals, i.e. a weaker interfering signal is suppressed in the receiver. The degree to which a particular receiver shows this effect is termed its *capture ratio* and is measured in decibels (dB). The FM transmitter is more efficient as the amplitude of the carrier is constant. This means that all RF amplifying stages can be operated at maximum efficiency.

Disadvantages of FM include the requirement of wider bandwidth than DSBSC if the signal-to-noise is maximized, and the capture effect being undesirable if it is the weaker signal to be received.

As the acquisition of the weaker signal is often required in aviation applications it is found that AM systems are often preferred in aircraft and air traffic control centres.

Phase modulation (PM)

Frequency modulation produces a phase displacement of the carrier, the peak displacement in radians being equal to the phase modulation index.

The only difference between phase and frequency modulation is therefore that, in phase modulation the phase displacement is proportional to the modulating voltage, and in frequency modulation the frequency deviation is proportional to the modulating voltage. Since the phase modulation radians equals the modulation index:

$$\phi_d = \frac{\omega_d}{\omega_m}$$
$$\omega_d = \phi_d \times \omega_m$$

The amount of frequency deviation is therefore directly proportional to the modulating frequency for PM, and the phase modulation is inversely proportional to frequency for FM. Phase modulation may therefore be produced on an FM system by increasing the modulating signal with frequency at 6 dB/octave. FM may be produced on a phase modulating system by decreasing the modulation signal at 6 dB/octave.

It is interesting to note that pre-emphasis at 6 dB/octave effectively changes the nature of an FM signal to PM at the higher modulating frequencies.

The means by which a transmitter produces FM or PM is dealt with in a later chapter. FM can be produced by changing the frequency of a free running oscillator, or with a synthesiser. A crystal oscillator, however, cannot easily be frequency modulated, and phase modulation is then applied. As several radians deviation may be required, this usually involves multiplication. The type of modulation produced therefore depends upon the master oscillator, but, as shown above, doctoring of the modulating system will convert FM to PM or PM to FM.

Pulse code modulation (PCM)

Modulation modes so far discussed affect the only two possible variables of the carrier – its amplitude or its frequency (phase). PCM is a mode of modulation which can be imposed on a carrier as either AM or FM. It does not, however, represent the information to be conveyed by the system directly, but is a representation of the speech converted into digital form. The advantages of this are that the message may be sent at much higher than 'real-time' speed, and utilising error correction techniques, a signal badly corrupted in transmission may be corrected in the receiver. A further advantage is that speech data in this form may be mixed with computer data to send a composite information and control signal. The prime disadvantage of this technique is the increased amount of circuitry required to implement the digital functions and the far tighter specification it imposes on the transmitting and receiving equipment. However, as VLSI devices become lower power and more system

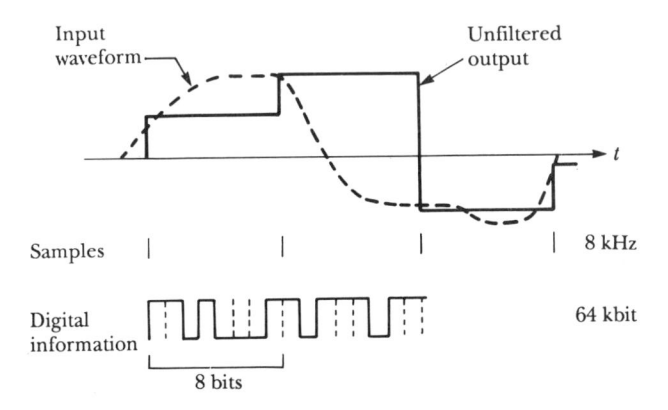

Figure 4.5 *Pulse code modulation*

orientated, future transmission modes will certainly become more digitized (see Chapters 15 and 16).

Figure 4.5 shows a wave converted to pulses as 'digital information' and also converted back again as 'unfiltered output'.

The speed at which we can transfer this PCM onto the carrier has limits (a function of system bandwidth) and the rate is referred to as the baud rate. The baud rate is a measurement of the number of bits (or pulses) which can be handled per second.

Waveform coding techniques such as PCM or delta modulation, require high data rates, typically 64 K/bits second, and would therefore be unsuitable for use in an all-digital radio system (poor spectrum/bandwidth utilization), hence the need for more efficient digitization techniques. The later section on Pan-European digital technolgy examines these techniques in more detail.

Receiver fundamentals

The principles discussed in this section apply to both AM and FM receivers. From earlier discussion, PMR has been defined as being more suitable to the VHF/UHF bands, 30–1000 MHz. The intelligence to be communicated has a frequency of 300 Hz to 3400 Hz. If 300 MHz is taken as an example of carrier frequency, and 3 kHz as the audio bandwidth, it is seen that this ratio is 100 000 to 1. To resolve such an extremely small portion of the RF spectrum in order to extract the modulation, the carrier must be processed with a high degree of accuracy and stability. A more practical approach is to change the carrier from its 300 MHz down to a lower frequency, but with the modulation unchanged. In this way a more manageable frequency is used from which to extract the modulation. The method used to convert the incoming RF carrier to a lower frequency is to mix this carrier with a second frequency which is generated within the receiver itself. The second frequency generated within the receiver is referred to as the local oscillator (LO). The circuit element which is used to mix or to beat the two signals (carrier and LO) together is called the mixer. The resulting lower frequency produced is known as the intermediate frequency (IF) because its frequency lies between the carrier frequency and the audio frequency. A receiver which uses this technique of beating (or heterodyning) two signals together is called a superheterodyne or superhet receiver. It is the receiver type which is currently used for all PMR work and the type which is discussed in the future chapters.

Figure 4.6 shows a typical superhet block diagram. The antenna is shown connected to a preselect filter stage. This stage matches the impedance at the base of the antenna into the receiver and provides some defining of the frequencies seen by the receiver – it has *selectivity*. The output is matched into the next stage, the RF amplifier. This stage amplifies the extremely small signal appearing at the output of the preselector (typically less than 1 μV) to a

Principles of RF communication

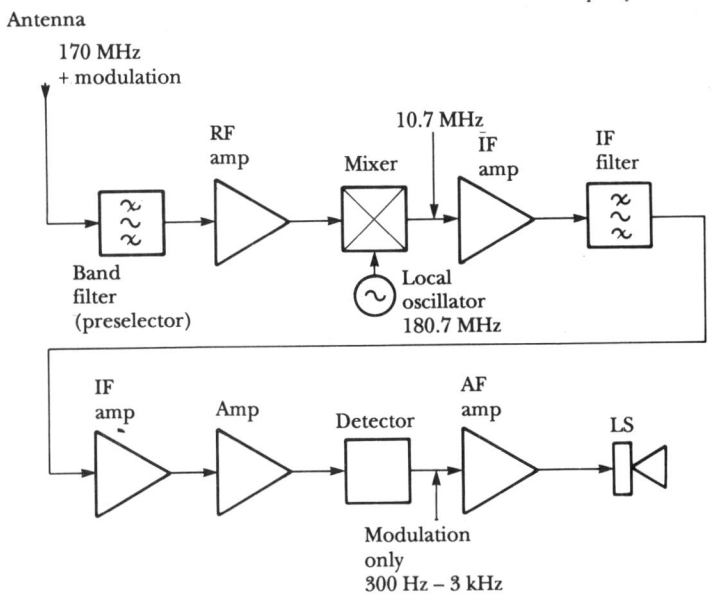

Figure 4.6 *Superhet receiver system*

level suitable for feeding into the next stage, the mixer. This device has two inputs:

1 the RF carrier from the preselector, and
2 the frequency produced by the LO.

The mixer will have at its output a frequency which is a function of 1 and 2 and this IF (typically between 400 kHz and 22 MHz) is matched into the next amplifying stage. Following this stage the signal passes to the IF filter. This filter is the prime component used to narrow the bandwidth down to the few kilohertz of audio. Note that the frequency on which the filter is centred is the wanted IF, but that its bandwidth is a few kilohertz wide. After the filter there are several stages of gain IF amplifiers to raise the signal to a level suitable for the modulation detector. The detector recovers the required audio frequency, passes it through an audio filter, and after audio amplification presents it to a loudspeaker or earpiece.

There is a second form of superhet which is often used, and this is the double superhet. In this receiver the mixer is followed usually by an IF amplifier but then the IF goes into a second mixer to be mixed with a second LO to produce a second IF. There is a filter at the first IF to reduce image problems, and then a second filter at the second IF of precise audio bandwidth. The advantage of this double superhet technique is that the first filter is required to be less precise than the single filter system and the second filter working at a very low IF (typically 455 kHz) is easier to produce and thus much cheaper.

Mobile Radio Servicing Handbook

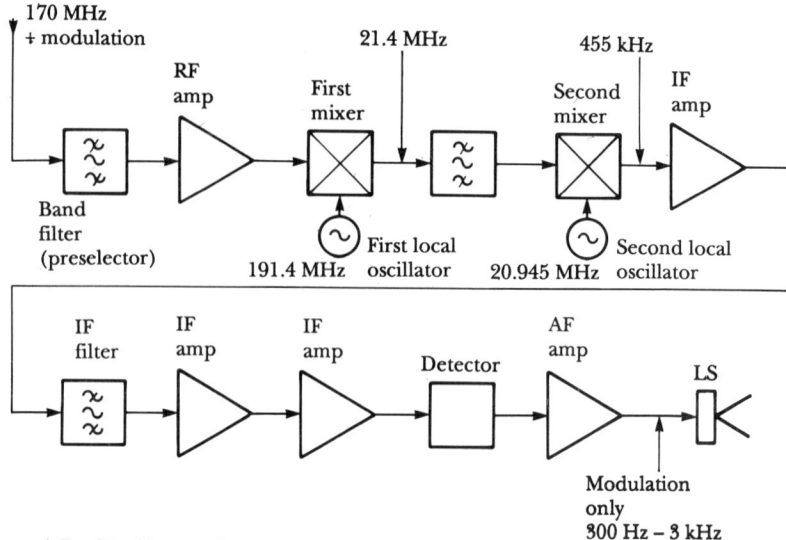

Figure 4.7 Double superhet receiver system

Transmitter fundamentals

The purpose of the transmitter is to produce the carrier signal at RF and with sufficient power for the system needs. This carrier must contain the modulated signal to be communicated, and it is the function of the transmitter to perform this modulation process. It may also be necessary for the transmitter to be able to change carrier frequency if it is acting in a multi-channel system.

At the heart of the transmitter is the generation of the carrier frequency to be transmitted. Various techniques are available to perform this function. A crystal may be used to generate a frequency which is a sub-multiple of the final carrier frequency. This oscillator is then followed by one or more multipliers to increase the frequency to the final requirement. The advantage is simplicity, the disadvantage is that the crystal is a single frequency device, and cannot be frequency modulated unless at small deviations.

A second method, and one being used increasingly in modern multi-channel equipments, is to use a frequency synthesizer. The frequency synthesizer uses only one crystal but has the ability to produce a large number of discrete output frequencies on demand. These frequencies, or channels, are at the final carrier frequency and are the same stability of output as seen with the simple crystal oscillator. The disadvantage of this technique is that it is considerably more complex in terms of circuit functions than simple oscillators, but modern VLSI is having a large impact in this area. The frequency synthesizer demands

Principles of RF communication

a higher level of knowledge from the service engineer and also some awareness of digital technology.

In the case of the AM transmitter, a sub-multiple of the final frequency is produced by crystal and then subsequent stages will both multiply the frequency and provide some power increase through to the final PA. AM is applied at the final PA, thus overcoming the need for power hungry linear stages through the transmitter chain.

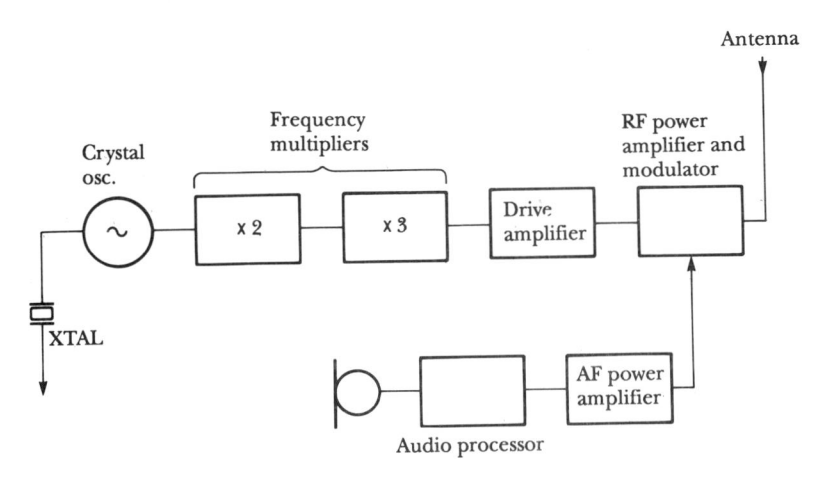

Figure 4.8 *Typical AM transmitter*

If the AM modulation takes place at the output of the transmitter at high RF power, then the modulating signal must be amplified in high power Class B* circuits to drive the high level modulator. However, the stages between the low

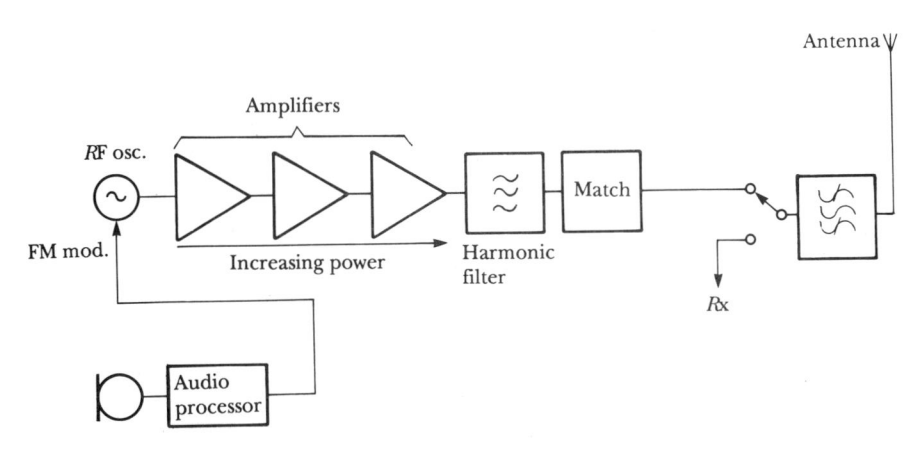

Figure 4.9 *Typical FM transmitter*

37

level oscillator output and the modulator (running at final power), can be efficient Class C* stages as there is no modulation envelope at this point.

In the case of FM the modulation is invariably applied to the frequency generator at the heart of the transmitter and as there is no amplitude varying envelope, all power stages are operated in Class C.

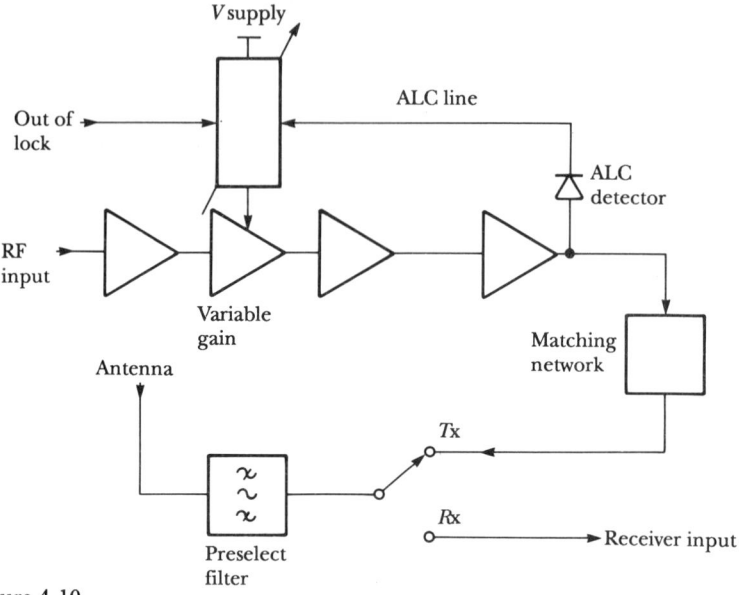

Figure 4.10

The output of the transmitter is matched into the antenna section by means of one or more stages of tuned circuits (using only passive components). This circuit both matches the impedance of the antenna to the output of the transmitter – usually an impedance transformation and also provides selectivity to assist harmonic suppression. In practice this matching circuit is often the same preselect filter used at the front of the receiver. A method of RF output power stabilisation under varying operating conditions will be included, as will a method of turning off the carrier if not on frequency.

The other section to consider in the transmitter is the speech or audio processing circuits. If normal speech is converted to an electrical signal and then analysed, it will be seen to consist of an average level of power interspaced with peaks of signal.

The mean power level of speech is of the order of 30 per cent. This means that if no alteration is made to the speech waveform, the modulator and RF circuits must be capable of handling the occasional power peaks without distortion and the mean level of utilised transmitted RF power is low. In order

---

* For definition of A, B and C circuits see Appendix II.

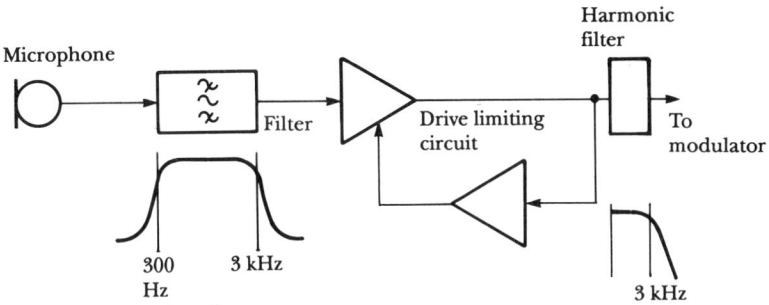

Figure 4.11 *Typical audio processor compressor*

to improve the efficiency of this situation, the audio signal is processed to increase its mean power level.

5 Practice of RF communication

Introduction

In Chapter 4, the basic requirements of an RF communication system were reviewed. The reasoning behind these requirements were considered and the advantages and disadvantages of alternative systems were discussed.

This chapter examines in some detail how the different methods and approaches already discussed are implemented in practice. This leads to a comprehensive build up of a complete receiving system, considering the many different features required to fulfill the system needs. The same will be done for the transmission system. Consideration will also be given to the various signalling methods which exist – single tone, multiple tone, DTMF etc. The various circuit elements used to generate and process these tones will be outlined.

Antennas and matching

The earlier superhet block diagram (Figure 4.6) showed various sections in a typical receiving system. The first part in the receiving system is the antenna.

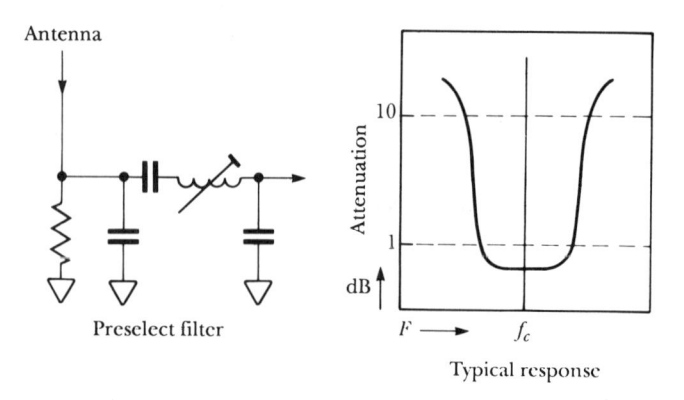

Figure 5.1 *Typical receiving system*

The PMR systems operate in narrow bands (typically less than 10 per cent of carrier frequency) in the VHF/UHF spectrum. The dimensions of an antenna are directly related to the frequency at which it is to operate. This means that the antenna has its maximum sensitivity across the band in which it is wished to operate, and becomes increasingly less sensitive at frequencies further away from this band. This helps to reduce unwanted signals and out of band noise from entering the receiver. A fairly typical type of antenna for hand-held PMR equipment would be a coiled helix having low impedance at its base and fairly omni-directional characteristics.

Matching and preselection

The antenna has a voltage generated across it by the wave which it is receiving. It is now necessary to transfer that voltage into the receiver signal circuits with as little loss as possible. This is achieved by matching the impedance at the bottom of the antenna to the first receiver filter section. This is done by means of a tuned network, typically a pi section network, which makes an impedance transformer. The purpose of the preselector filter is to provide a 'clear path' response only for those frequencies in the band to be received.

With the amount of radio communications these days, there is always a large amount of transmitted RF power, both in the operating band and adjacent to it. It is important to prevent as much unwanted signal as possible from entering the active circuits of the receiver and causing unwanted effects. The preselector filter will usually consist of a number of tuned pi section filters optimised for a flat response in the passband and as great a rejection of frequencies either side of this as possible (Figure 5.1). Note that the last section of the filter will also perform a matching function into the following circuit.

RF amplification

Because the antennas used on portable and semi-portable equipment are small, only a small voltage is developed across them. Also, as the preselect filter is composed of only passive circuit elements it has a loss figure, even in the passband. Hence, before the RF signal can be processed it is necessary to increase its amplitude. The prime purpose of this is to lift the level of the signal out of the residual (or unavoidable) noise generated in the mixer of the receiver. The signal is fed to the RF amplifier. This amplifier has two main requirements in this circuit position:

1 low noise performance, and
2 large signal handling capability.

It is clear that the function of the receiver is to produce the wanted signal at the output in a clearly intelligible form. If the signal heard contains a high level of random noise it will be difficult to resolve. As the ratio of noise to signal worsens, so a practical limit of receiver operation is reached. All stages in the receiver will generate this unwanted or random thermal noise. By careful selection of materials in the semiconductors used and by well designed circuits, it is possible to minimise it, but not eliminate it. In the later stages of the receiver – IF, AF etc. – the wanted signal is large, i.e. millivolts or volts. The addition of a few microvolts of noise to this signal will go unnoticed. The problem occurs in the first stages of the receiver, the RF amplifier and the mixer. At this point the signal itself is only a few microvolts, so the addition of a small amount of noise will have a catastrophic effect.

It can be seen that the ability to reproduce clearly very low level signals (the definition of *sensitivity*) is dependent on the amount of noise generated in the first stages of the receiver, and also losses in the antenna circuits. This highlights the necessity for the RF amplifier to have an excellent noise performance. There is however a second requirement for this amplifier, large signal handling capability. This may at first seem a strange requirement. However, all the signals received in the operating band should be considered. The strengths of these signals will vary from that of powerful local base stations to weak distant hand-helds. The function of the RF amplifier (or any amplifier) is to increase the amplitude of the signal applied to its input. A signal is applied to the input and the gain of the amplifier causes a larger signal to be seen at the output. If the input signal continues to be increased a point is reached at which the amplifier is unable to provide sufficient output signal for the required gain. If the amplifier has, for example, a 20 dB gain, all signals fed into the input will experience this gain. When however the point is reached at which the output

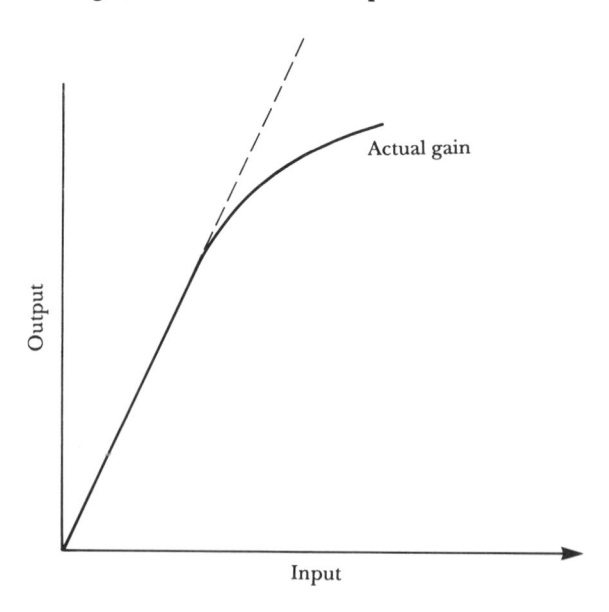

Figure 5.2

Practice of RF communication

begins to limit, the gain starts to fall. The amplifier is said to be in a state of gain compression (Figure 5.2).

This means that if a very high level signal is fed to the input, the amplifier may provide only 17 dB of gain. A further increase in signal input may well reduce this to 16 dB. So, as the amplifier sees strong input signals its gain falls. This may not be a problem apart from the distortion produced if the strong signal is required, however it may be a very low level signal which is required in the band. If the gain of the amplifier is 20 dB the small signal will be amplified by this amount.

The gain of 20 dB will be enough to lift the signal well above the noise level or threshold of the amplifier. The problem occurs when the low level signal sees only, for instance, 14 dB gain. Whilst this compression effect is occurring, the noise generated within the amplifier remains constant and so the signal-to-noise ratio or sensitivity of the receiver is considerably reduced.

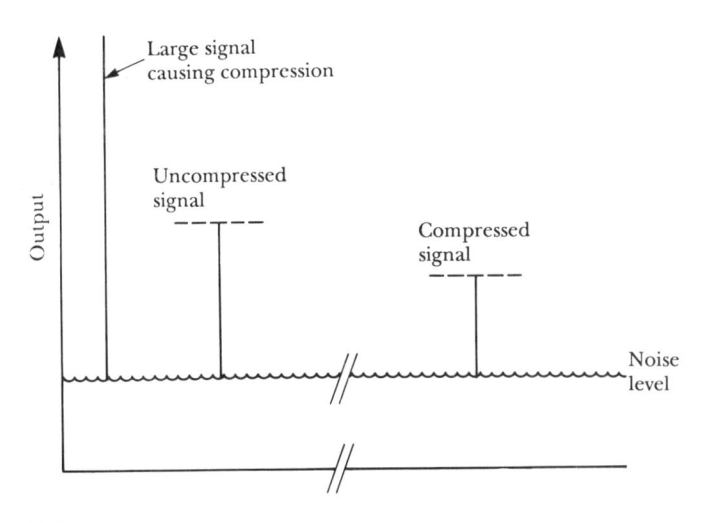

Figure 5.3

A second problem can occur under these conditions. As the peak power of the large unwanted input signal rises and falls with modulation, so the gain of the RF amplifier will rise and fall. This effect will cause the amplitude of the wanted low level signal to rise and fall in sympathy, i.e. modulation. The operator will find himself listening to the modulation of the unwanted signal. This effect is known as *cross modulation*, for obvious reasons.

The effect is essentially a problem for an AM interfering signal and an AM receiver. With an FM receiver, cross modulation from a strong AM signal is largely suppressed by subsequent limiting, and only the residual phase modulation caused by the interfering signal causes trouble.

The signal level handling capability of an amplifier determines the point at

Mobile Radio Servicing Handbook

which this happens. The cross modulation effect is independent of the desired signal level, but is proportional to the square of the undesired signal amplitude. Because of this relationship it is a function of the signal handling capacity for the amount of RF attenuation produced at the front end of the receiver. Since both cross modulation and compression are caused by strong undesired signals which are adjacent to the receiver passband, they can be controlled to a certain extent by the preselection filter, which has already been discussed.

However, if the signals, the strong and the weak, are all in band, they can only be controlled by the amplifier having a large signal handling capacity so that this compression does not occur within the operating range of the receiver.

It has been seen that the RF amplifier requires both low noise performance and large signal handling capability. These two parameters do not go hand in hand. For large signal handling, a fairly high level of current through the front end amplifier is required. This in turn generates noise, so the design is a compromise and requires careful consideration.

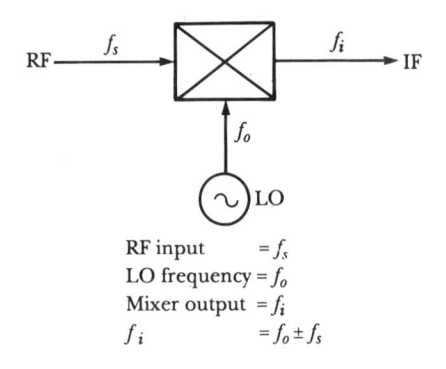

RF input $= f_s$
LO frequency $= f_o$
Mixer output $= f_i$
$f_i \quad\quad = f_o \pm f_s$

Figure 5.4 *Mixing*

Mixing

Following RF amplification the signal needs to be 'translated' down to the lower intermediate frequency (IF) prior to detection. The circuit element to do this is the mixer. The output of the RF amplifier is fed to one input of the mixer via a matching circuit. The other input of the mixer is fed by the local oscillator (LO). The mixer will take these inputs and 'beat' the two frequencies together to produce frequency products. The ideal mixer has a square law characteristic (i.e. the transfer function has no higher terms than second order) and, under such conditions, would produce just the sum and difference frequencies of the input, i.e.

RF Input (F1) = 170 MHz
LO Input (F2) = 159.3 MHz

The output = F1 $\pm$ F2

Output 1 = F1 + F2 = 329.3 MHz
Output 2 = F1 − F2 = 10.7 MHz

Output 2 is chosen as the IF as it is considerably lower than the RF input, and hence easier to process. As the mixer is not an ideal device but has some unwanted non-linearity, other unwanted frequency products occur and it is these that the IF filter rejects.

If the LO is 159.3 MHz, there is a second input frequency with which it can mix to give 10.7 MHz, i.e. 148.6 MHz. As the receiver IF cannot know which of the two 10.7 MHz signals are valid, the only way to overcome this problem is to ensure that the input frequency of 148.6 MHz does not reach the mixer. This can only be achieved by front end selectivity. This unwanted signal is called the *image frequency* because it is a 'mirror' of the wanted signal. It is apparent that the image frequency is removed from the required frequency by twice the IF, and the higher the IF, the easier it is to suppress the image frequency.

The sum and difference equations can be satisfied by injecting the LO on either the high side of input signal or the low side. In low band PMR receivers the LO is injected on the high side of the carrier so that, for a given operating frequency band, the LO has less percentage range to cover, thus easing the oscillator design.

In high band PMR receivers the LO is injected on the low side of the carrier so that the image frequency lies on the high side where the RF amplifier has a falling gain (frequency response). The preselect filter used to block the image frequency will have a steeper response on its high side, helping to overcome this problem.

If, because of the fixed nature of the following filters the IF is always 10.7 MHz, it is seen that to bring a different input frequency into the IF passband it is necessary to change the LO frequency. In other words the receiver is tuned by changing the frequency of the LO. This tuning range is restricted only by the frequency range available from the LO and the RF preselect filter bandwidth.

It is seen that as well as the wanted IF at 10.7 MHz, there are unwanted frequencies produced in the mixer. The following stages have the task of narrowing down the frequency band to be processed so that these signals are not passed through the receiver to the detector.

Note: for an RF input frequency f_1, and a local oscillator f_2, ideal modulation will produce the product:

$$A \cos 2\pi f_1 t (\cos 2\pi f_2 t) = \frac{A}{2} \cos 2\pi (f_1 + f_2)t + \frac{A}{2} \cos 2\pi (f_1 - f_2)t$$

All practical modulators produce additional outputs, since they do not operate as perfect multipliers. Additional outputs include $f_1, f_2, f_1 + 2f_2, f_1 -$

Mobile Radio Servicing Handbook

$2f_2, f_1 + 3f_2, f_1 - 3f_2$ etc. The type of mixer and the amplitude of f_2 affect the ratio of wanted to unwanted output.

IF filtering

The input to this section is a band of signals centred on the IF frequency, in this example 10.7 MHz. This, as yet unfiltered, IF band may be several MHz wide. As the bandwidth required is only 3 to 4 kHz or 10–12 kHz wide for AM and FM respectively (centred on 10.7 MHz) a very high degree of selectivity is needed.

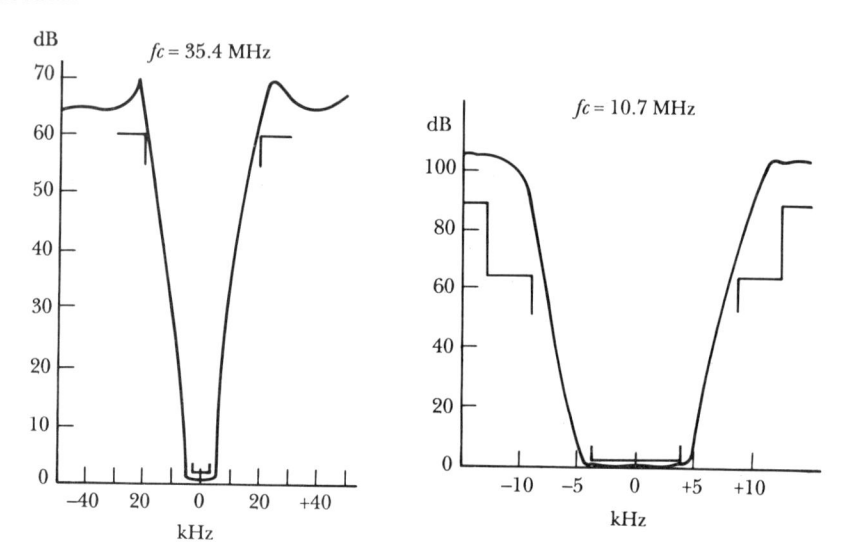

Figure 5.5

This selectivity cannot be practically obtained from a simple capacitive inductive filter with repeatable stability and low cost construction. A different approach must be used. In this circuit position a ceramic or crystal filter is used, either of these having sufficient selectivity to define the narrow audio bandwidth. As the output from the mixer and input to the filter are generally very different impedances, a matching network is required. This usually consists of a simple inductive capacitive filter stage, typically a PI or L network. The stage following the filter will be an amplifier with a very high input impedance such as a Mosfet.

This technique is used so that the output of the filter may be impedance matched with a simple passive component network, making the matching independent of the in-band signal level. This network will have a variable component, usually a capacitive trimmer, in order that the filter impedance may be matched exactly. Mismatch of the filter causes not only loss of signal

Practice of RF communication

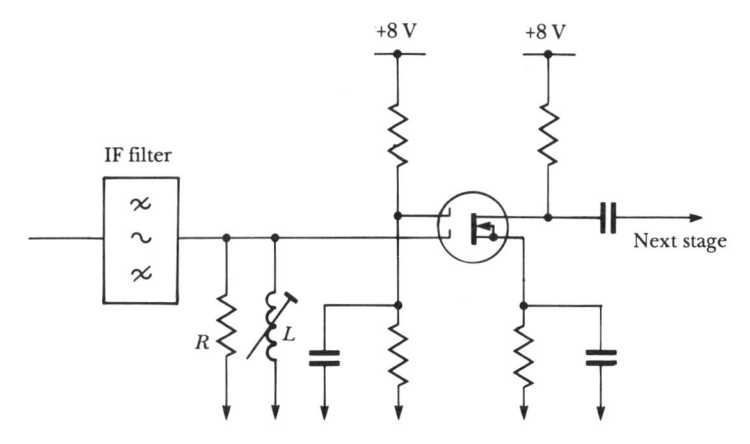

Figure 5.6

but also a differential delay across the band of frequencies passed through the filter. Although this may not be noticeable in the recovered audio, it can cause very high error rates if the receiver is used to recover digital data due to a 'skewing' of the bit stream.

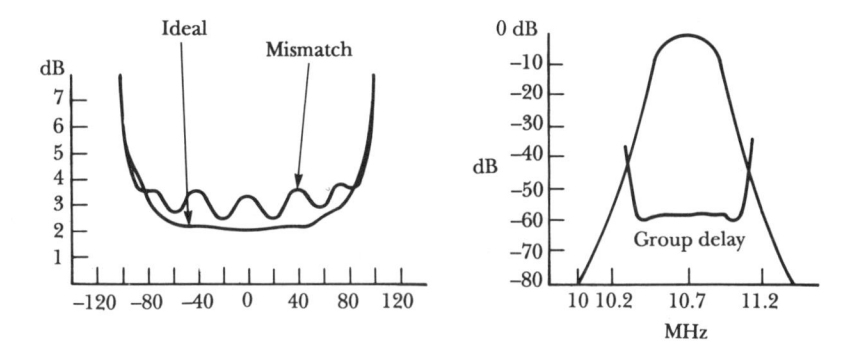

Figure 5.7

In a double conversion superhet system the first IF filter will not provide the total selectivity required. Its purpose will be to remove noise and unwanted mixer products at the first IF. As it has a broad response and rolls off well outside the audio bandwidth, it is often referred to as a 'roofing' filter. This filter output (after matching) will be fed to a second mixer which will have a second locally generated frequency mixed with this first IF, e.g. 10.7 MHz. The product of the first IF and the second LO will be a second IF typically between 100 and 500 kHz. At this second IF, it is quite simple to provide the degree of selectivity required with a simple LC network. Note that the tuning of the receiver still takes place at the first IF by varying the first LO. The

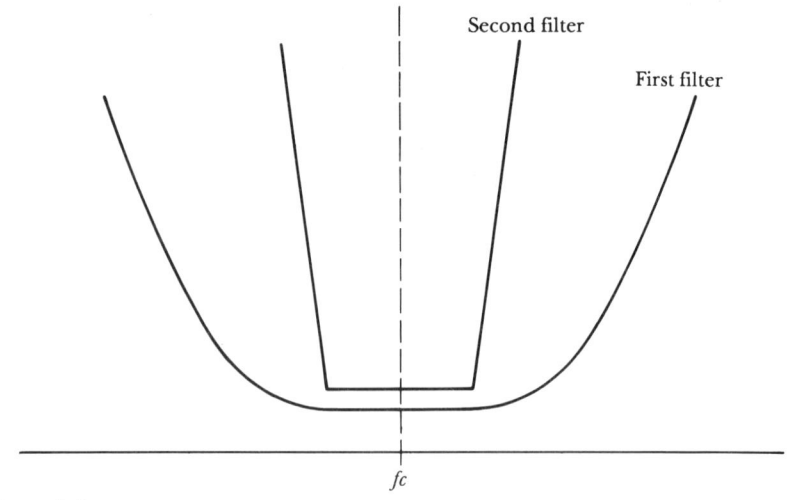

Figure 5.8

second LO is required only to convert a fixed first IF to a fixed second IF and can hence be a fixed crystal oscillator.

Until this point, the circuits and processes described are common for both AM and FM receivers. It is now that the two modes require different processing.

IF amplification and detection for AM

The signal from the mixer is in the order of a few millivolts, and after the loss in the IF filter, may well be in the microvolts region. This is not sufficient to drive an AM detector and so a large amount of gain is necessary in the IF strip. This is produced in either discrete bipolar transistor or field-effect transistor (FET) stages, or in an integrated circuit. The IF filter may also be placed midway in the strip so that its severe out of band attenuation reduces the noise from the lower level IF gain stages. One or more IF amplifiers will have a second input, and by means of a varying DC voltage on this second input, their gain may be controlled. This requirement is discussed further under the automatic gain control (AGC) heading. The output of the IF gain strip is fed to the AM detector.

In the AM system the modulation signal varies the amplitude of the RF carrier. The superhet principle converts this RF carrier down to an IF still with the AM present. It is now necessary to recover this amplitude variation as an electrical signal. The usual method for this is to rectify the IF. The output of the diode rectifier will be a voltage varying as the amplitude of the IF. As the amplitude of the IF in the modulation troughs is very low, it is necessary to use a diode with a low forward voltage drop. A germanium diode or hot carrier diode is frequently used. A second AM detector method is sometimes used and this is the transistor detector. In this case the IF gain amplifier and transistor detector is included in one integrated circuit.

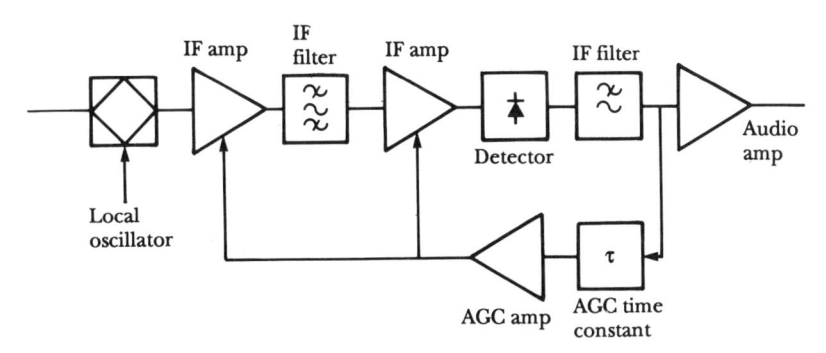

Figure 5.9

AM automatic gain control AGC

As the received signal strength varies over quite a wide range (with varying alignment of antennas, local signal shadows and buildings) some method is required of keeping the signal presented to the diode detector at a constant amplitude. An 'automatic gain control' system can be used and by rectifying the IF, a voltage is obtained that is proportional to the carrier strength.

This rectified IF, a DC voltage, is directly proportional to the strength of the received signal. This voltage can be used to vary the gain of earlier IF amplifiers so that when a fall in signal strength is detected the IF gain is increased and when an increase in signal strength is observed the IF gain is decreased. The timing of this action is a subject of critical design, as if it is too quick in its effect it will tend to follow the low frequency AM, and by gain variation of the amplifiers remove it. If on the other hand it is too long, the effect for which it is compensating will have gone and when the IF amplifiers are being instructed to increase their gain the signal may well be getting stronger. The result is that the whole AM IF section oscillates or is, at least ineffective.

IF amplification and detection for FM

The signal for the mixer is at a low level (mV) and although the mixer will provide gain, this is offset to some extent by the loss in the IF filter. This means the signal is not sufficent to drive an FM detector and so gain is required if the IF strip. Unlike the AM system, it is not necessary to maintain the correct amplitude of the signal at all times but only its instantaneous frequency. This makes the FM IF strip simpler in that the AGC system is not necessary, just a chain of amplifiers providing gain sufficent for

Mobile Radio Servicing Handbook

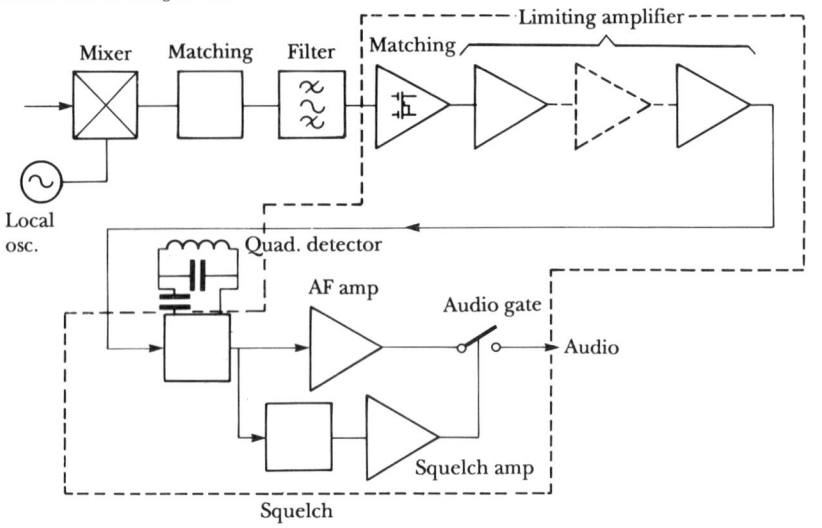

Figure 5.10

the final amplifier to limit at the lowest level of signal input. This means in spite of varying input levels there will be a constant level out of the IF strip. This is known as limiting and, in addition to providing an AGC action, it removes any amplitude modulation existing on the signal, and also suppresses spike interference.

This constant level signal is fed to an FM detector or discriminator. This circuit element takes the IF input and produces an output whose amplitude is directly proportional to the deviation of the IF input, and whose frequency is equal to the modulating frequency. The two common passive types of discriminator, the Foster Seeley and the Ratio Detector, both use tuned circuits and diodes to perform this function. The Foster Seeley discriminator uses a tuned-transformer primary coupled to a centre tapped tuned secondary circuit. The rectifiers are orientated in a bi-phase configuration. The result is a lead-lag of phase about the tuned centre frequency giving a linear output with FM.

The Ratio Detector is so called because it relies on the ratio of two voltages developed from the input signal rather than absolute levels. It uses a tuned primary circuit inductively coupled to a tuned secondary circuit and a third seperate winding used to generate the phased ratio action. The output characteristic is similar to the Foster Seeley. The Ratio Detector is the more popular because it has limiting action and operates over a wide signal level. Both types of discriminator, however, require careful alignment to ensure linearity and maximum sensitivity. Other types of discriminators to be found are a digital type known as a pulse counting discriminator or a system utilising a phase locked loop.

The most common type of FM discriminator used for both wide band (broadcast) and narrow band (PMR) is the quadrature detector. This type invariably occurs in IC form as all sections apart from the quadrature coil can be

Practice of RF communication

invariably occurs in IC form as all sections apart from the quadrature coil can be included on the chip. Modern chips include the limiting IF amplifiers, demodulators, squelch circuits, AF pre-amplification and often a mixer with oscillator for superhet receiver use. The detector uses a linear phase shift network tuned to the IF producing a 90° phase shifted signal. The direct IF signal is multiplied by the phase shifted signal in a circuit similar to a balanced mixer. The output will be a series of pulses having a time duration related to the phase differences of the input signals. The output is then filtered to produce a smoother voltage following the pulse duration (the instantaneous frequency of the IF), and hence the original modulation is recovered.

De-emphasis as described in the previous chapter, is achieved after FM demodulation by means of a simple RC (or LC) network to attenuate the higher frequencies. The network will produce a roll-off rate of 6dB/octave for PMR systems.

At this point the recovered modulation may be 'picked off' and fed to CTCSS or SELCALL decoder circuits for further processing.

Squelch and audio output

As discussed in Chapter 4, when the signal level at the antenna falls, the signal-to-noise ratio at the output of the receiver decreases and the noise level in the output rises considerably. This situation is particularly bad in the case of FM, as it is always operating with full IF gain and the discriminator cannot distinguish between a valid signal and random noise in its operating band. It is thus necessary

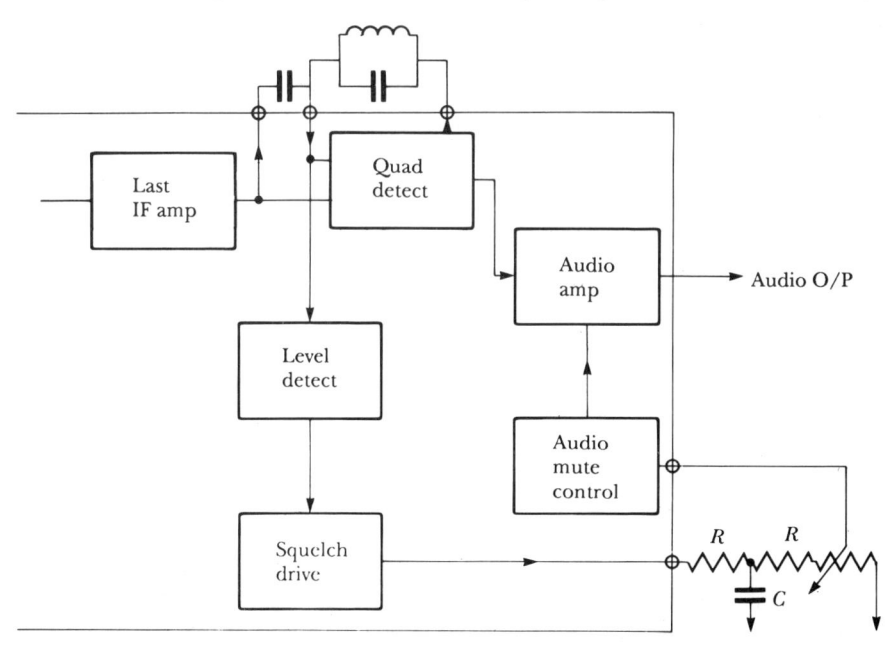

Figure 5.11 *Audio squelch*

Mobile Radio Servicing Handbook

to have a means of detecting this high noise condition and to take action to prevent it from reaching the loudspeaker. The noise generated in the receiver circuits and detected by the discriminator has a frequency content from the lowest audio frequencies up to the maximum response of the discriminator, usually in excess of 10 kHz. When a valid audio signal of sufficient strength is being discriminated, this noise is suppressed and only the transmitted audio frequencies exist at the output of the discriminator. From the above considerations it can then be seen that if there is a circuit at the output of the discriminator which will only respond to frequencies well in excess of the audio band (300–3400 Hz), it will provide a means of detecting the low signal high noise condition.

The circuit consists of a simple high pass filter which will only pass signals of greater than typically 6/7 kHz and a buffer amplifier. This amplifier is usually a saturating amplifier such that its output will switch hard on or off depending on whether a signal is present or absent at its input. The output of this amplifier will be used to operate a switch (usually a transistor) in the audio path, thus enabling or disabling the audio signal route to the output. By having a control which can set the sensitivity threshold of the amplifier the operator can determine the noise level at which the audio is turned off. This function is known as squelch or muting.

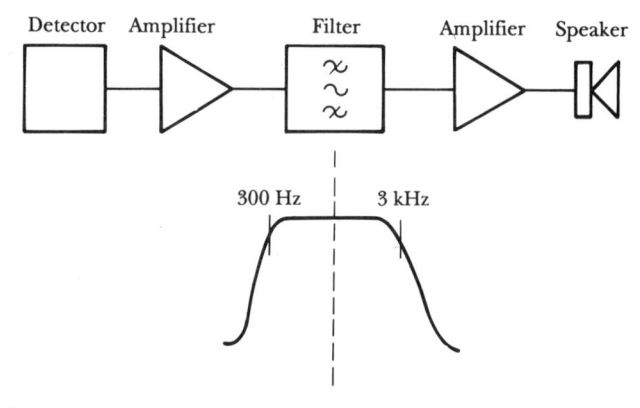

Figure 5.12

Audio output

As speech frequencies are contained in the band 300 – 3400 Hz, it is undesirable to have frequencies outside this range present in the loudspeaker. They will usually consist of noise, distortion or any other unwanted products generated within the receiver. For this reason a sharp bandpass filter is employed in the audio output stages. To obtain sufficient rejection of unwanted frequencies and to prevent loss in the passband, it is an active filter configuration. This is usually an emitter follower circuit or, in more modern equipment, a four element (or quad) operational amplifier configured as a high pass low pass filter assembly.

6 Practice of RF communication – transmitter design

Transmitter design

A prime function of the transmitter is to generate the RF carrier frequency to be transmitted. This will be achieved either by means of a crystal oscillator or by frequency synthesis. See Chapter 7 for a detailed consideration of oscillator requirements and synthesizer action.

The generated carrier must also be modulated with the audio signal to form the composite wave for transmission.

Audio processing

Examining a typical speech waveform when it has been converted to an electrical signal on an oscilloscope shows it to have certain marked characteristics.

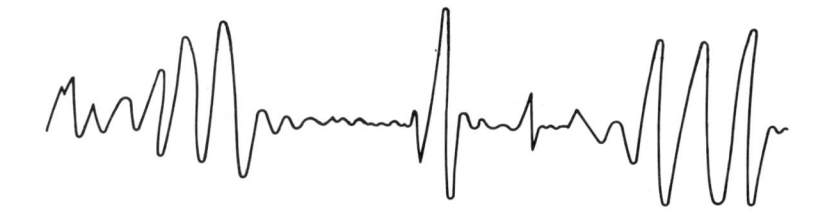

Figure 6.1 *Typical speech waveform*

The waveform, indeed any waveform, can be shown to be made up of a number of sine waves added together. The frequency content of the waveform is expressed in terms of the minimum and maximum frequency of sine waves needed to synthesize this waveform. In the case of the speech waveform, it is found that using only frequencies between 300 Hz and 3400 Hz, it is possible to synthesize or 'put together' any speech patterns or waveforms with no loss in intelligibility. This means that the audio circuits can be profiled to have a passband response only over these frequencies and all high frequency and unwanted background noise will be eliminated from the output. This comment applies to both the output circuits of the receiver and the microphone input circuits of the transmitter. If a passage of speech is created – a few words – and then the waveform captured on a storage oscilloscope, other characteristics can be examined. The speech will be seen to consist of mainly a similar level of signal, an absence of signal between words and occasionally during words, large peaks of signal. This means that certain sounds made in speaking require a peak of sound or sound energy. If the waveform is analysed further it is found that the 'normal' level is about 30 per cent of the peaks.

In the FM transmitter the operational range achievable is not only a function of the RF power transmitted, but also a function of the deviation, i.e. the power in the sidebands. It is this fact that produces the tendency to overmodulate. This must be avoided as the effects of overmodulation are severe, including generation of RF harmonics and splatter into adjacent bands. If the audio waveform previously described is to be handled by later stages of the transmitter, they must be capable of handling the peaks of this waveform without distortion. However, by far the largest proportion of the speech signal is 70 per cent below this level. The situation is even worse with AM. If the carrier is not suppressed, two-thirds of the transmitted power is wasted in the carrier, even at 100 per cent modulation. If normal speech level is equivalent to 30 per cent modulation, the intelligence in the transmission only accounts for about one-thirtieth of the transmitted power.

This is the reason why the modulation power is often adjusted to too high a level without considering that voice peaks overmodulate the transmitter and cause a certain amount of splatter. There are, however, various means of increasing the mean depth of modulation without increasing the peak modulation level. When used correctly they allow a noticeable increase in volume and thus coverage without causing splatter. There are two methods in popular use:

Dynamic compression The mean depth of modulation can be increased using the so-called dynamic compressor. Such circuits use the AGC principle where a control voltage is produced from the output level of the AF amplifier and used to control the path gain so that the output peak signal is virtually constant. The problem with these circuits is generally the same as the AGC problem described in the section on receivers, i.e. the time constants have to be very carefully designed so that the gain does not rise between words and hence produce a great amount of background noise as well as echo effects. On the other hand the rise time constants should not be small enough to control the lowest modulation frequency transmitted. For this reason the dynamic compressor is often used in conjunction with an audio filter, and so provides a fairly complex, but effective circuit.

Amplitude limiting (clipping) In this mode the audio amplitude on peaks is limited

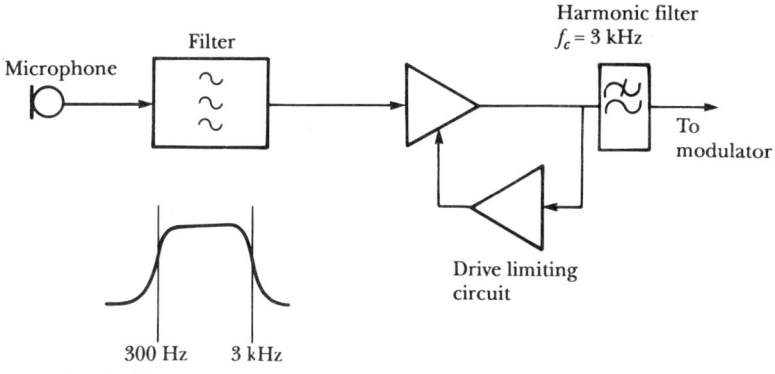

Practice of RF communication – transmitter design

Figure 6.2 *Audio processor*

to a certain value. This process is often called *clipping*. Since the voltage peaks which were responsible for the over modulation of the transmitter are eliminated in this manner, it is possible for the modulation level to be increased so that the limited voltage modulates the transmitter to a depth of 100 per cent. The more the voltage peaks are limited the higher will be the mean depth of modulation. The limit to this process, however, is the distortion which will occur as clipping becomes more and more severe, and under such conditions the limited signal will contain a large number of audible harmonics which will cause the modulated RF signal to be increased in bandwidth. Present generation equipment tends to use the audio clipping or limiting process to an extent before these undesirable effects occur. However, some sophisticated equipment uses a process by which the clipping occurs at RF level. The audio is mixed up to RF, clipped and then reconverted back to audio frequencies. The advantage is that all unwanted frequencies lie well outside the audio band and so are easily removed. This is not yet in common use but, due to the higher level of integration of many equipments, it may well appear more in the future.

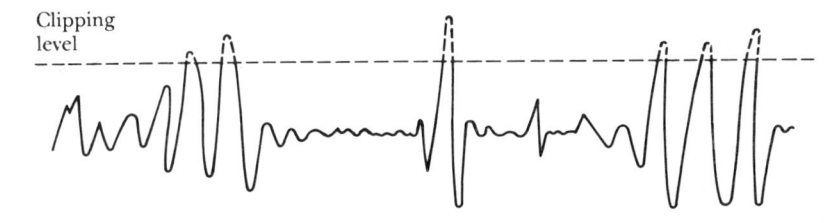

Figure 6.3 *Amplitude clipping*

Pre-emphasis, as previously described, is also achieved in the audio processor circuits. A simple frequency conscious network (usually a CR combination) will be inserted to roll the higher audio frequencies up at a rate of 6 dB/octave.

RF signal generation

A quartz crystal will be the basis of any RF generation because the allowable tolerance of the frequency is very tight – of the order of a few parts per million (ppm). A number of factors contribute to frequency error, but two most potent ones are:

1 temperature variation of the crystal itself;
2 ageing of the crystal.

In the 1960s and 1970s it was common practice to fit crystal elements into temperature controlled ovens which held them at an elevated temperature (usually between 70 and 80°C), and the crystal elements were cut such that their frequency variation with temperature curves were flattest here. As the ambient temperature fluctuated over −10 to +55°C, the crystal temperature would typically be held say within ±5°C. Then came a period where crystal polishing techniques enabled a tolerance of ±5 and even ±3 ppm (by selecting crystals) to be achieved over −10 to +55°C which met the UK DTI specifications on all bands without the use of an oven.

Now however, the DTI specifications call for tighter tolerance than this, particularly with base station equipment which has called for the re-introduction of the oven, since ±3 ppm over the full temperature range is approaching the economic crystal limit. Oven technology has also improved and now they mostly consist of a plastic power transistor to perform the heating, in conjunction with a PTC resistor in the circuit as a sense element. They are now cheap enough to consider fitting to every base station. Ovens of ten years ago used a wire element with a thermostat and were notoriously unreliable.

Ageing is an effect within the crystal element where the impurities within the sealed can get absorbed onto the quartz slice, making it more massive and hence oscillate more slowly. The effect is exponential, i.e. during the first year, the total ageing (in ppm) would be the same as the next two years put together, which is the same as the next four years put together and so on. Crystals which have a temperature tolerance of ±5 ppm over the full temperature range must have an ageing rate of at least one tenth of this upon installation to avoid many frequency adjustments early on. Thus, more crystals this tight have to be pre-aged at manufacture for perhaps five days, which adds even more to their cost.

A crystal which is designed to run in an oven and is optimised for operation at 80°C, will have a tight temperature tolerance (±2 ppm) over say 5°C either side of 80°C, but can be relatively poor (±10 ppm) outside this range, keeping the cost of the crystal down. The crystal ageing rate will be accelerated at an elevated temperature so it is important to pre-age the crystal which tends to offset the saving. However, when one considers that ten crystals may be discarded to find one with the required ±3 ppm over the full temperature range, the most economical and best technical solution is a modern ovened crystal if ±3 ppm or better is required.

Practice of RF communication – transmitter design

Some ageing is inevitable and will require visits to the transmitter (and receiver for that matter) to adjust the frequency. It is suggested that visits are made after 1 month, then after 2 months, 4 months etc., on an exponential basis, after which only annual visits are required.

The RF signal is generated by an amplifier with the crystal in its feedback path, causing the amplifier to oscillate at the crystal resonant frequency. The nature of resonance is discussed in Chapter 7. The circuit configuration is therefore known as a crystal oscillator and will be referred to as such.

The crystal then is the source of accurate frequency, whether the oscillator circuit containing the crystal is directly or indirectly used, providing the distinction between non-synthesized and synthesized signal sources. Frequency synthesis involves using a second oscillator (not with a crystal) which is locked to the crystal oscillator with a phaselock loop and is the subject of the next chapter. A non-synthesized signal source uses the crystal oscillator output directly.

In theory crystal oscillators can run at any frequency, but in practice they are restricted to between about 1 MHz and 20 MHz; below 1 MHz the size of the crystal element becomes too large, above 20 MHz the crystal element becomes impractically thin and delicate, with a high loss. The upper limit of 20 MHz refers to the crystal element operating in its fundamental mode, i.e. the crystal element is stressed with one half cycle, much like a string vibrating up and down. However, with a string it is possible to vibrate it such that an exact number of half cycles can fit between its ends. This is when the string or crystal is oscillating at an 'overtone' – if five half-cycles are present, that is the 5th overtone frequency, and this enables a crystal oscillator to be built to run at frequencies above 100 MHz.

Modulating the RF signal

The RF signal source is thus generated, and after perhaps some buffering by another amplifier, it has to be processed. According to whether the transmitter is synthesized or not, or is AM or FM, the RF signal is processed differently. In the case of AM, whether synthesized or not, the modulation is put onto the RF signal very late on at the power amplifier stages just prior to the antenna circuits. With AM synthesized transmitters, the synthesizer output is usually

Table 6.1 Transmitter configurations

Type	RF multiplication	Modulation used
AM synthesized	Yes ($\times 2$ usually)	AM at PA stage
AM not synthesized	Yes ($\times 2$ to $\times 12$)	AM at PA stage
FM synthesized	No	FM at 2nd oscillator
FM not synthesized	Yes ($\times 2$ to $\times 36$)	PM after 1st oscillator

PM = phase modulation.

doubled in frequency to overcome modulation crosstalk problems. FM synthesized transmitters put the modulation onto the RF signal by interfering with the second non-crystal oscillator frequency, i.e. within the synthesizer, making it vary according to the amplitude of the AF signal. FM transmitters using crystal oscillators directly use phase modulation (explained later in Chapter 8), since the crystal oscillator cannot easily be pulled off frequency as discussed earlier in this chapter. All transmitters using crystal oscillators directly will use frequency multiplication stages after the oscillator, whether or not an overtone crystal is used. In summary, see Table 6.1.

The effect of frequency modulation may be obtained by phase modulation with an audio signal which has been processed to attenuate at 6 dB per octave. The reason phase modulation is used with a direct crystal oscillator is because frequency modulation cannot be used as the crystal oscillator will not move in frequency (which is what is required). The oscillator stage is therefore followed with a phase modulator.

Amplification

The RF signal as generated is very low in power, typically 10 mW, and without amplification will not propagate very far. The power must be boosted to perhaps several tens of watts before applying it to an antenna. FM synthesized and non-synthesized transmitters and AM transmitters all have different RF processing.

Non-synthesized transmitters (transmitters which directly use the crystal oscillator as their signal source), and AM synthesized transmitters need to multiply the RF signal frequency as well as increase its amplitude. The two functions are normally produced together with medium power amplifiers. The amplifiers are arranged in a chain, each one using higher power transistors than the stage before. Early on the amplifiers are overdriven such that they operate in a non-linear region and so produce harmonics which are, of course, exact multiples of the RF at their inputs. Filters at their outputs then choose one of the harmonics and reject all others, such that a non-linear amplifier plus a filter can be considered as a multiplier stage. It is quite common to have ×2 and ×3 multiplier stages in the correct cascade configuration to obtain the overall desired multiplication factor which might be anywhere between ×2 and ×36. It is most unusual to have the last two or three amplifier stages in the transmitter performing any multiplying, as the efficiency of a multiplier stage is well below that of a straight amplifier. If the transmitter power is 25 watts, then the final frequency would have been reached by the stage operating at, say, 1 watt.

With FM synthesized transmitters, the second (non-crystal) oscillator will be the one whose output is amplified, and this oscillator will run on the correct transmitter output frequency making multipliers unnecessary. Multipliers are inefficient and quite noisy, so getting rid of them has advantages. However, in contrast to a transmitter with multipliers, one without them necessarily has a lot of amplifiers all working at the same frequency. This causes the designers considerable problems with stability.

In the case of normal AM (DSB plus carrier), once the amplitude of the RF signal is correct for the transmitter output the modulation is applied. The output amplitude is varied up and down in sympathy with the audio waveform by altering the current passing through the RF power amplifier transistor. This is normally achieved by passing the DC to this stage through a transformer secondary winding. The primary winding is driven by audio power transistors with the audio signal. Approximately half the RF power is required of the audio amplifier to achieve 100 per cent modulation; if the RF power is 25 watts, then some 12.5 watts of audio power is necessary into the modulation transformer to modulate the RF signal adequately. Thus, AM transmitters waste power.

With the modulated RF signal at the correct amplitude, however it is generated and modulated, it must be applied to an antenna without losing any power in the process. The signal is taken through a matching circuit to bring the impedance to 50 ohms, which is used universally on mobile radio transmitters. The signal is then taken through some switching circuits in the case of a transceiver, since during receive the antenna must obviously be connected to the receiver. The final stage is a filter which removes all the harmonics of the RF signal produced by the power amplifier stage. The harmonics are produced because this stage will be operating in Class C which takes it into non-linearity, and whereas with multiplier stages this was used to advantage, here any harmonics are regarded as spurious emissions and have the potential to interfere with other users.

On FM transmitters there will be a power detector on the transmitter output whose output is amplified and fed back to a driver stage in conjunction with a potentiometer to adjust and control the output power; this is known as an ALC loop. AM transmitters do not have these as it upsets their modulation characteristics; instead they usually vary the RF coupling in the later amplifier stages to obtain different output powers.

The references above to AM apply to normal double sideband with transmitted carrier, and the limited use of AM for commercial mobile purposes is at present restricted to this simple, although inefficient method of transmission. As explained on page 28, the full advantage of suppression is only obtained by suppressing both carrier and one sideband (SSBC). The circuitry for suppression operates on the low level signal. The amplifier stage can then operate at lower power, since the carrier has been suppressed, but the amplifier stages need to be linear to avoid distortion of the modulation.

7 Frequency synthesizer circuit principles

Introduction

Until recently PMR systems were limited to operation on two or three channels only, unless the operator was prepared to open the equipment and change plug-in components. With the rapid increase in communication requirements a method was needed that would both pack more channels into a given frequency band and make it easy to access these channels. Packing more channels into a given band has meant that channel spacing is getting narrower, 6.25 kHz being the aiming point of present systems. It is obviously desirable that PMR equipment manufacturers can produce and sell a range of equipment for use on any channel group or spacing without the need to 'customize' the sets.

Clearly there is a need for a circuit module which performs as an oscillator, gives multiple frequency outputs on choice and is cheap and simple to produce. Such a module is the frequency synthesizer and so important is it to new and future equipments that service engineers must have an understanding of its operation.

We will look first at where oscillators are used in PMR systems and hence define the characteristics required by these applications.

RF oscillator – theory and practice

RF communications operate on the principle of a transmitted wave containing the information (speech) to be sent to the recipient. This wave will have a defined frequency dependent on the allocation of a suitable channel. To allow for the amount of communications traffic, as many channels as possible are allocated in a particular band of frequencies. This means if carriers are to be placed close together (12.5 kHz or 6.25 kHz), good frequency stability of this carrier is required. It can be seen from this that the transmitter has the requirement to generate a number of stable signals at the channel operating frequencies.

To return to the principle of receiver tuning (Chapter 5) the superhet system was discussed. A key part in this frequency conversion process is the mixer. The mixer has two inputs:

Frequency synthesizer circuit principles

1 The band of signal frequencies received by the antenna, passed through the preselect filter and then amplified.
2 A signal generated within the receiver itself at a chosen frequency.

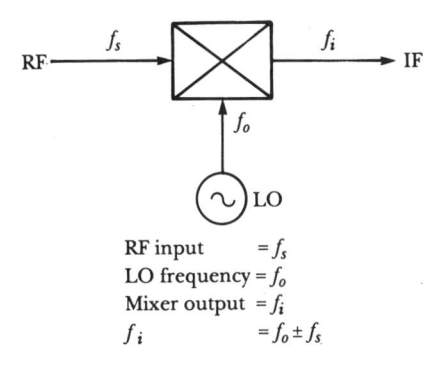

RF input $= f_s$
LO frequency $= f_o$
Mixer output $= f_i$
$f_i \quad = f_o \pm f_s$

Figure 7.1

The other element of importance in receiver tuning is the IF filter. This was described as an electromechanical component, either quartz or piezoceramic having a fixed centre frequency and a well defined passband of a few kHz.

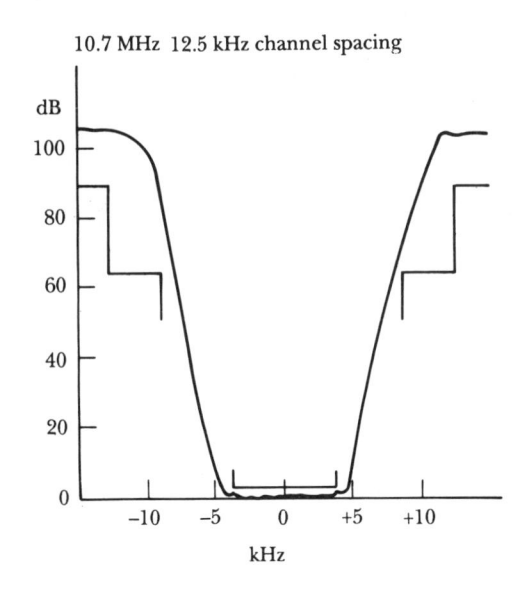

10.7 MHz 12.5 kHz channel spacing

Figure 7.2

From consideration of these points it can be seen how a particular frequency is selected for reception.

The IF is fixed at 10.7 MHz by the IF filter. So:

61

Mobile Radio Servicing Handbook

If the local oscillator (LO) = 159.3 MHz
IF = 10.7 MHz
then the received frequency = 159.3 + 10.7
= 170.0 MHz

As the IF is fixed at 10.7 MHz, the only change possible is to the LO frequency. It can be seen what happens when a change is made.

If the LO is now made 159.4 MHz

the received frequency = 159.4 + 10.7
= 170.1 MHz

so it is seen that the receiver is retuned by 100 MHz.

By changing the LO frequency, the receiver is tuned, as long as the received frequency is within the preselector filter passband. This requirement demonstrates the need for generating a large number of defined stable frequencies to be used as the local oscillator to tune the receiver to its many channels. This need is the same as the requirement for the generated carrier in the transmitter. There is, however, one important difference. In the transmitter all frequencies are generated directly on the channel frequency, in the receiver all frequencies are offset by an amount equal to the IF.

In Chapter 5 the parameters influencing receiver sensitivity were discussed and it was seen that it is primarily a function of noise in the front end of the receiver. The mixer was included in the definition of the front end (as at this stage the signal level is still extremely low) and so inputs to this section must be noise free if maximum sensitivity is to be obtained. This means the local oscillator must be low noise for maximum performance. We can now see how a crystal oscillator matches these ideal requirements.

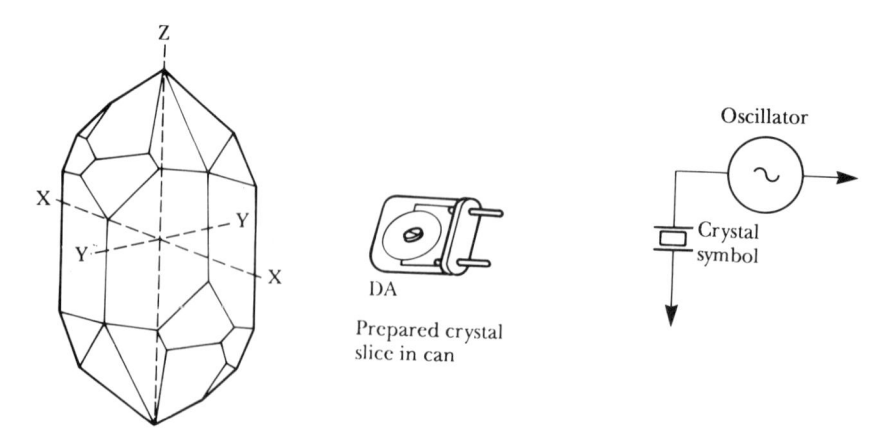

Figure 7.3 *Quartz crystal*

Frequency synthesizer circuit principles

A crystal is a block of naturally occuring mineral called quartz. Quartz exhibits a property known as the Piezo electric effect, which means that it will change shape under the influence of an electric field. The thickness of a slice will determine its natural resonant frequency and this can be stimulated by electrically exciting the electrodes deposited on opposite faces of the slab.

To fix the frequency at which the crystal vibrates or resonates, it is cut into a thin disc. The dimensions of this disc determine precisely the frequency of oscillation. This means that as crystals approach resonant frequencies of only tens of kilohertz, they become large and relatively expensive. At the upper end of frequency performance (about 25 MHz) they become extremely thin and hence difficult to process in production. To protect them in use they are sealed in a metal can filled with an inert gas (or dry air) to give constant performance. To set them resonating an electrical field is applied across the disc of quartz to excite the crystal lattice into vibration. If a crystal is stimulated to resonate at the frequency at which one half wavelength fits between the faces of the slab, it is said to be operating in its fundamental mode. It is possible, however, to force it to resonate at a number of half wavelengths and hence the oscillator will produce multiples of the fundamental frequency. Such a crystal is said to be operating in overtone mode and the crystal in the way that it is cut, will be characterized for this operation at production.

The crystal element can be made to run on an overtone of its fundamental frequency; overtone crystals are made to run at up to 200 MHz or more. Overtone crystals are electrically more robust in terms of frequency tolerance, that is the frequency of the oscillator depends less on the components external to the overtone crystal than with the fundamental crystal, and the higher the overtone the better they are in this respect. The reasons are outside the scope of this book, but it explains why, for example, an 18 MHz crystal may in fact be a third overtone 6 MHz crystal, rather than a fundamental 18 MHz crystal. The crystal, of course, is happy to operate on its fundamental frequency as well as on the first few overtones; higher than about the seventh overtone the crystal becomes quite lossy and difficult to drive, and in practice the highest used is third or fifth. It is up to the circuit designer to ensure that the amplifier has enough selectivity and gain to oscillate only around the desired crystal overtone. For example, 12 MHz crystal operating at seventh overtone will produce a frequency of 12 × 7 MHz, i.e. 84 MHz. In practice this could then be doubled in a following circuit and a frequency of 168 MHz would be obtained.

If operation on more than one channel is required, it would need more than one crystal, a clearly impractical solution if operation is needed on more than three or four channels. However, certain characteristics of a crystal oscillator are unique and fulfill part of the ideal oscillator requirement.

An oscillator can be considered as a tuned circuit coupled to an amplifier so that when certain feedback criteria are met, the circuit oscillates. The frequency is determined by the inductance and the capacitance – a change of either produces a change of output frequency.

Figure 7.4

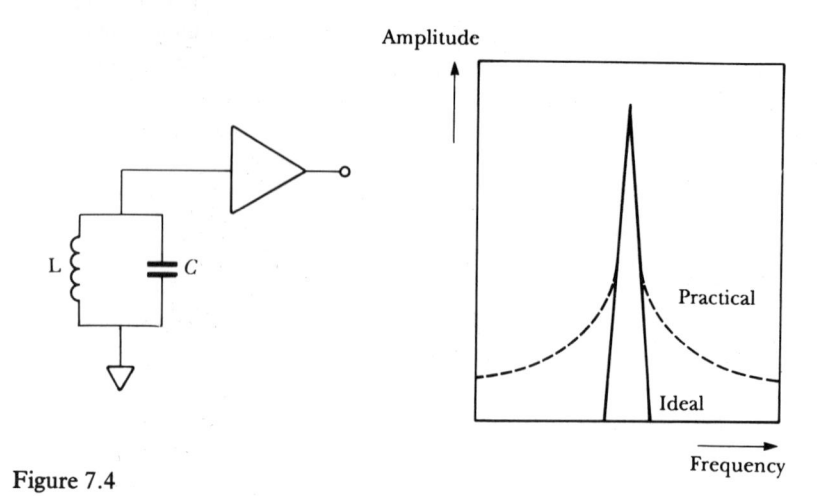

An 'Ideal' oscillator for radio applications would produce a single frequency with no noise associated with it. In practice, if the oscillator output is examined in detail, i.e. using a spectrum analyser, noise or random frequency effects will be observed close (kHz) to the 'single' main frequency. If this noise is fed into the mixer of the receiver it will degrade the sensitivity.

The noise is a direct function of the 'Q' or purity of the tuned circuit. The

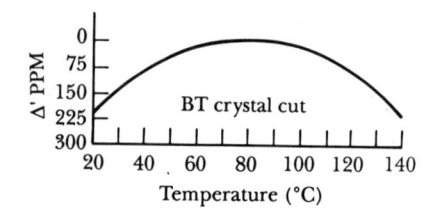

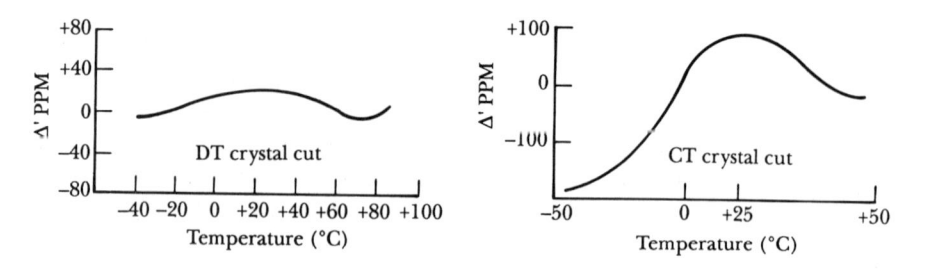

Figure 7.5

Frequency synthesizer circuit principles

ideal can be approached in the crystal oscillator where the 'Q' can be up to 1×10^6, but in the case of the LC oscillator several hundred is considered good. Stability of the crystal oscillator is also extremely good, it has less temperature dependence than the LC circuit. As noise decreases with increasing Q, the crystal oscillator obviously has a purer or more noise free output than that of the LC oscillator.

The second problem is stability. A change of either the inductance L or the capacitance C will produce a change of frequency, and heating and cooling effects within the equipment will certainly change the dimensions, and hence the values of the L and the C. This effect is considerably smaller in the case of the thermally stable quartz crystal.

The advantage of the LC oscillator is that it is comparatively easy to change the L or the C as required. In older equipments the capacitor was made variable by means of sets of intermeshing plates fixed to a rotary spindle. By rotation of this spindle a greater or lesser area of plates meshed and hence the capacitance was varied. This gave the ability to change the oscillator frequency on an infinitely variable basis. Thermal instability was, however, still a problem.

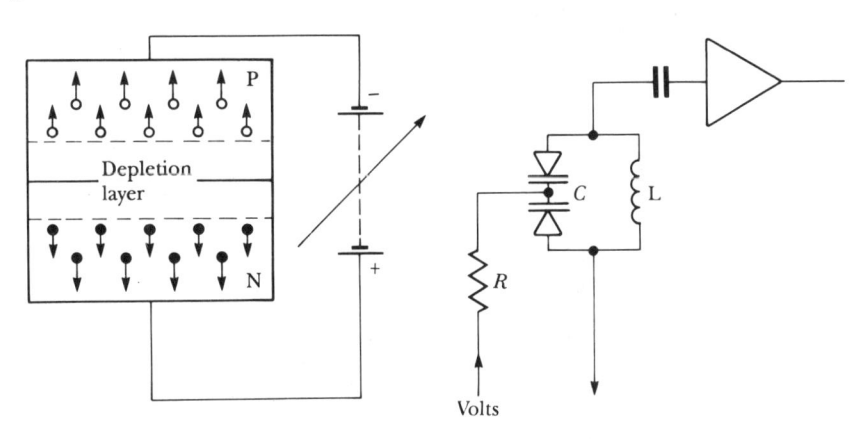

Figure 7.6 *The variable capacitance diode*

The modern counterpart to this is the variable capacitance diode or varicap. This diode is composed of two semiconductor materials – a p or positively charged material, and an n or negatively charged material. When these two materials are fused together, an electrical barrier is set up either side of the junction. At the junction, a re-combination of the charged particles takes place, leaving an electrically charged gradient on either side. Because the area either side of the junction has no charged particles, it is called a depletion layer. This depletion layer is analogous to the dielectric of a capacitor and the charged areas either side to the plates. The varicap diode can be used as a capacitor in the oscillator circuit where the capacitance and hence the frequency is

Mobile Radio Servicing Handbook

dependent on the width of the depletion layer. As the depletion layer is an electrically dependent region, its width can be made greater by a higher potential difference across the junction. Hence, by applying a greater or lesser DC voltage in reverse bias across the varicap, the depletion layer can be made wider or narrower. This means the capacitance exhibited by the diode is decreased or increased. If the varicap diode is now used in the oscillator circuit it can be seen that by varying the DC potential to the diode, the frequency of the oscillator will be varied. Note: Two varicaps are used to minimize harmonic generation due to forward diode biasing and to obtain a larger capacitance change with voltage. The problem of thermal instability still exists due to the temperature dependence of both the inductor and the varicap capacitor.

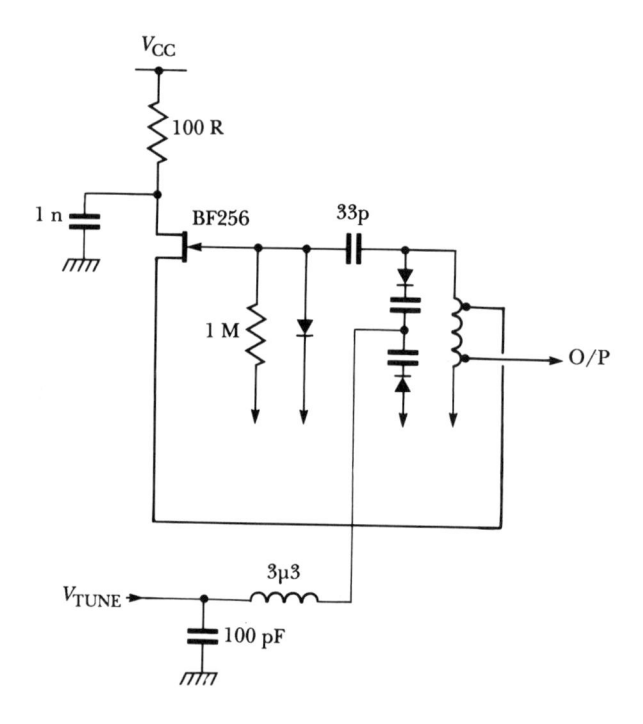

Figure 7.7

To summarize, the crystal oscillator provides a highly stable low noise output, but is inflexible in choice of frequency while, on the other hand, the varicap oscillator provides flexibility of frequency choice but is unstable in its output due to thermal and mechanical shock effects on its inductor and thermal effects on the varicap capacitor. The aim therefore is to bring the benefits of both types of oscillator together.

Before a complete system is discussed, two further circuit functions used in the phase locked loop must be examined

Frequency synthesizer circuit principles

Frequency and phase comparators

In a frequency synthesizer this block is usually referred to as the phase comparator. The phase comparator compares the frequency and the phase of two signals presented to its inputs and produces either of two outputs:

1. An output to indicate that the frequency or phase of input 1 leads the frequency or phase of input 2.
2. An output to indicate that the frequency or phase of input 2 leads the frequency or phase of input 1.

If inputs are of the same frequency and phase, it will produce an output that is midway between output condition 1 and output condition 2.

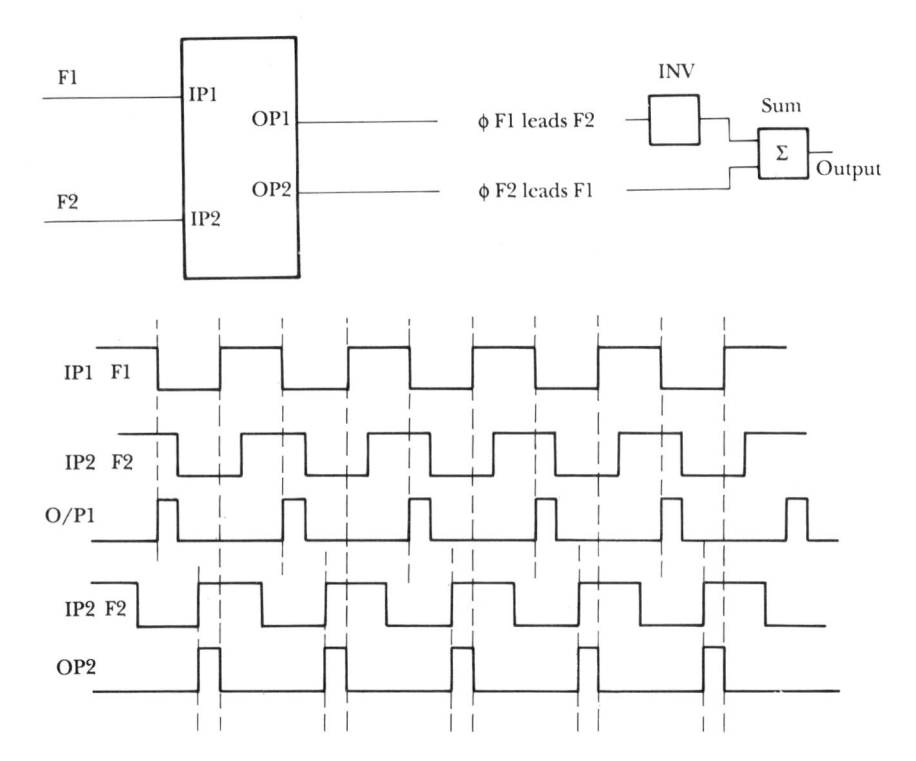

Figure 7.8 *Phase comparator and waveforms*

The outputs take the form of pulses whose amplitude is identical and constant, but where the difference in width is indicative of the frequency or phase difference between the two inputs. Figure 7.8 shows the effect of F1 leading F2 and the effect when F1 lags F2. One output is now inverted and summed with the remaining one. The phase comparator now has a single output. As the pulse width of the two outputs previously referred to is

67

identical, where there is no frequency or phase difference between the two inputs, this inversion and summation will produce an exact cancellation and hence there will be no output from the single output point. If however a phase or frequency difference exists between the two inputs, the single output will produce either negative or positive pulses, dependent on the sense of the difference.

In practice the ability to distinguish signals of different frequency and the ability to distinguish signals of different phase are often performed by two separate circuit blocks within the phase-frequency comparator, but having a common output. Note, therefore, that the phase detector gives both frequency error information where there is a frequency difference, or phase difference information where the two frequencies are equal.

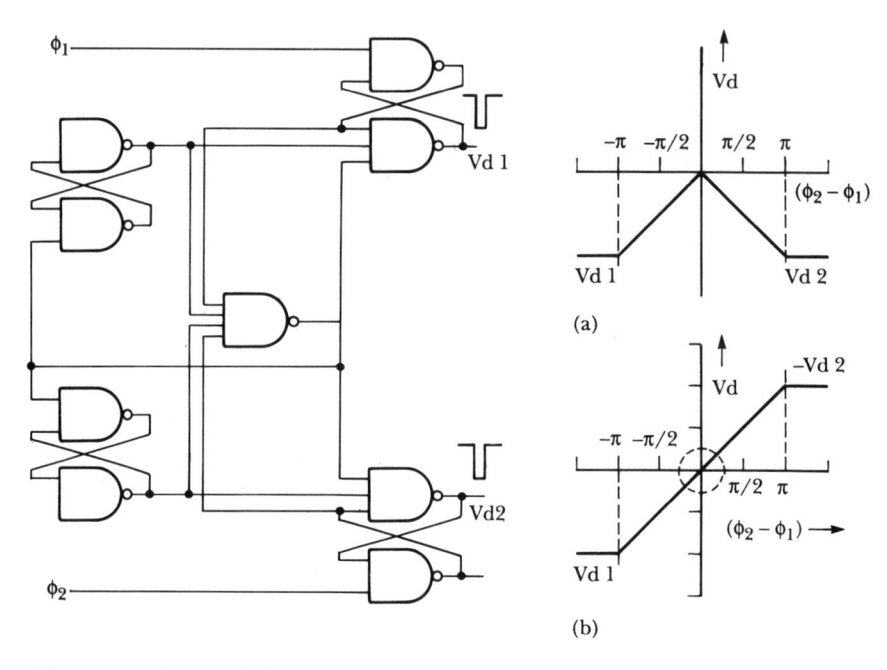

Figure 7.9 *Practical phase-frequency detector:* (a) single O/PS, (b) inverted/summed O/PS

The loop filter

The following table reviews the action of the phase comparator.

	Input phase	Output
1	f1 = f2	No output

Frequency synthesizer circuit principles

| 2 | f1 leads f2 | Positive going pulses |
| 3 | f1 lags f2 | Negative going pulses |

Note that for lines 2 and 3, the definition of which input sense produces positive and which produces negative pulses is arbitrary depending on the particular system design, i.e. line 2 could read

| 2 | f1 leads f2 | Negative going pulses |

and line 3 would be changed accordingly. The width of the pulses produced is directly proportional to the degree of phase difference between the two inputs, and the gain of the comparator circuit i.e. the greater the phase difference the greater the pulse width.

The output of the phase comparator is fed to a low pass filter. This filter has the action of smoothing the pulses presented to its input. If the filter has sufficient damping effect, i.e. it has a long enough time constant, it will remove completely the pulse action and produce at the output a DC voltage proportional to the width of pulses at the input. Remember, the pulse width is proportional to the phase difference at the inputs to the phase comparator, so this DC voltage produced at the output of the filter is a measure of this phase difference. The loop filter is an integrator with a zero and at least one pole. If it is a multiple pole filter, one will be dominant at the lower frequency with lesser poles at higher frequencies. The lower the frequency of this dominant pole, the slower the action of the filter.

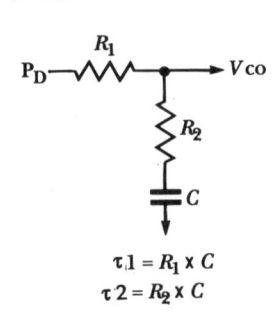

A typical passive loop filter. R_2 dictates δ.

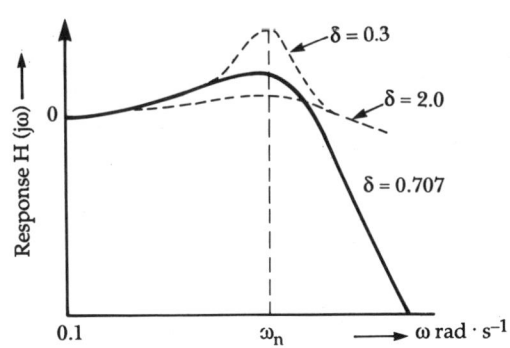

$\tau_1 = R_1 \times C$
$\tau_2 = R_2 \times C$

Frequency response of closed loop with values of damping factors (δ)

Figure 7.10

As most radios operate from a single supply, e.g. 12 V, no point in the circuit may go more negative than 0 V (unless a negative voltage supply is derived within the equipment). An indication is required of the phase difference and sense of the two phase comparator inputs, i.e. of the negative going pulses and of the positive going pulses. If the situation were adopted where f1 phase equals f2 phase is 0 V, then the filter output would need to go both negative and positive to follow the sense of f1 and f2, i.e. positive for line 2 of the table and negative for line 3. This is clearly inconvenient with single supply equipment. To overcome this problem a DC voltage derived from the supply rail is fed into the phase comparator/filter section. This voltage is in the order of a few volts. If f1 phase equals f2 phase then the filter output would be 4 V and would have the possibility to go up to +8 V for line 2 of the table and down to 1 V for line 3 of the table. Note that because of practical constraints of active circuit elements, this voltage does not go completely between 0 V and +12 V.

To summarize, in a typical system a 1 V output would indicate that f1 lagged f2 by a considerable amount. As the two phases were made similar the output would rise towards 4 V. When the output was exactly 4 V it would indicate that f1 and f2 were identical in phase. As the phase of f1 was increased relative to f2, the output voltage would rise towards the upper limit of 8 V. If the differences between f1 and f2 were greater than the phase comparator limits (see appendix) or f1 and f2 frequencies were dissimilar, then the filter output would be held against the limit of 1 V or 8 V.

The phase locked loop PLL

We can now put the circuit blocks which have been considered so far together as a system.

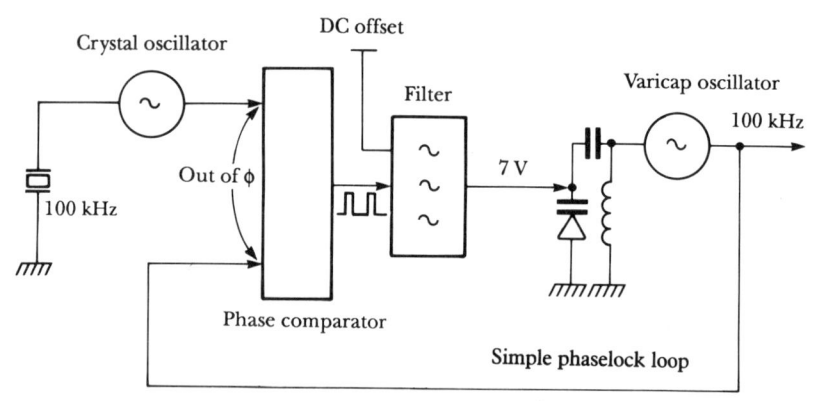

Figure 7.11

To set the conditions in the loop, we have a stable, noise free crystal oscillator running at 100 kHz. This is fed to one input of the phase comparator. There is a VCO whose frequency of oscillation is determined by the inductance L and the varicap diode capacitance C. The value of the inductor L is selected such that when combined with the value of capacitance resulting from 4 V DC applied to the varicap diode, the LC combination will result in an oscillator frequency of 100 kHz. The output of the variable capacitance oscillator or

Frequency synthesizer circuit principles

VCO is also fed to the second input of the phase/frequency comparator.

Consider now the loop condition. The phase comparator has two inputs:

1 the fixed 100 kHz reference frequency, and
2 the 100 kHz from the VCO.

Note as long as the two inputs are in phase using the example quoted earlier in this chapter, the DC voltage applied to the varicap diode will be 4 V and this is the voltage needed by the VCO to produce 100 kHz. So the loop is locked in a stable state by the phase comparator – a phase locked loop.

If now the inductance changes due to vibration or a change in temperature or some other effect, the frequency of the oscillator will start to move to a new frequency. However, this change in frequency will be fed to the phase comparator so that the phase relationship between input 1 and input 2 will start to change. This change in phase relationship will alter the pulse width out of the phase comparator, the filter will smooth these pulses and the DC voltage will start to change. This changing DC voltage will alter the capacitance of the varicap which will in turn bring the frequency of oscillation back to the 100 kHz. The oscillator will once again be running at 100 kHz but with a different value of capacitance to correct for the change in inductance, and a small phase error to produce the required correction voltage.

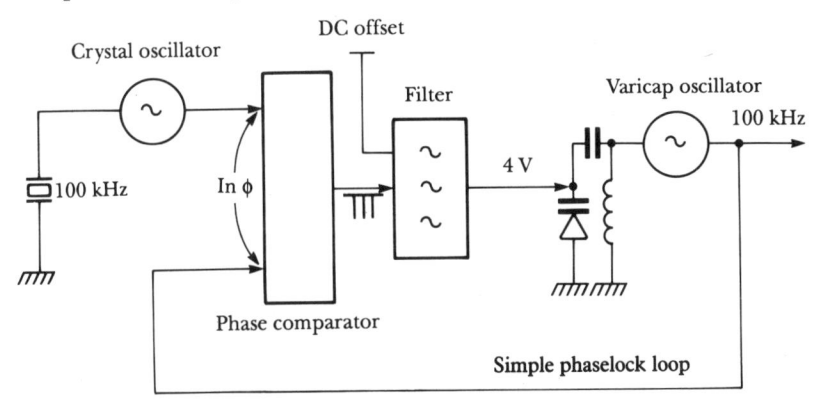

Figure 7.12

There are two points of which to be aware:

1 The loop can only correct for effects which take the DC correction voltage between the limits of 1 V and 8 V (using previous example). If the inductance, hence frequency, change is greater than this, the capacitance cannot be corrected further and the frequency will sit above or below 100 kHz, depending on the inductance.
2 The speed with which corrections are made is dependent on the loop bandwidth. The loop bandwidth is primarily a function of the loop filter.

The loop filter time constant is critical in correct loop action. Too much filter

Mobile Radio Servicing Handbook

time constant and the reaction of the loop is very slow. Similar to a pendulum the correction voltage will build up such inertia that it goes through the correct settling point and out the other side. If the time constant is very long it will never achieve stability but oscillate slowly about the correction point. Too little time constant and, although the loop will lock very quickly, the pulses from the phase comparator output still remain on the control line. These pulses will cause the varicap diode to change its capacitance to follow them, hence the oscillator output will be very noisy and jittery. As has been seen, this degrades the receiver performance and gives a noisy transmitter output.

Now the 100 kHz crystal is keeping a 100 kHz VFO stable. It would seem to be a lot simpler to throw away all the loop elements and use the crystal oscillator output directly. It is the next step which shows the benefit of the loop.

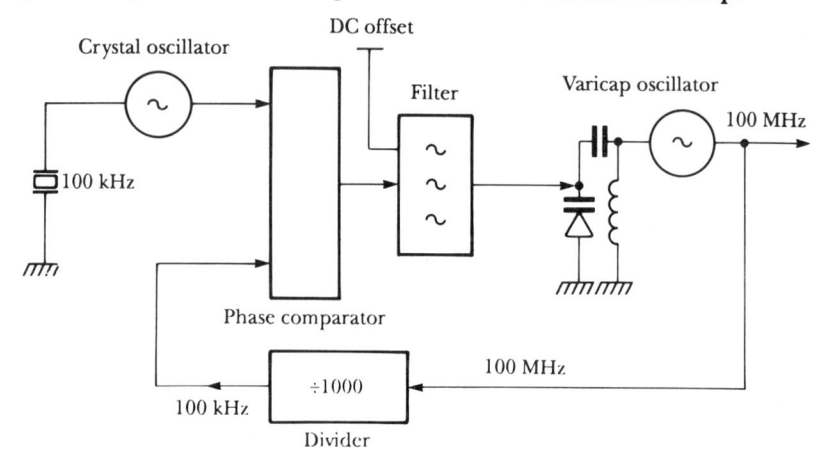

Figure 7.13

Figure 7.13 shows the previous loop but now with one new element added. This element is a divider, a digital divider. It has the ability to take in a particular frequency and divide it by the number which is set in this divider, i.e. if input to this divider is 100 MHz and it is set to divide by 10, 10 MHz is the output from the divider. If the divider is programmed to divide by 100 and is fed with 100 MHz, then 1 MHz is the output from the divider. As the diagram shows a division ratio of 1000 set into the divider, it can be seen that the divider output will be 100 kHz into the phase comparator.

The loop will sit in a stable, controlling condition whilst the variable frequency oscillator is 100 MHz, the divider is set to divide by 1000, and the reference frequency is 100 kHz. So now the stable crystal at 100 kHz holds the VFO at 100 MHz.

The next step is to see how to use this principle to give the ability to select a number of discrete frequencies.

The new element in the loop is a digital divider. If by means of control inputs, the division ratio is changed to 999, what occurs in the loop can be observed.

Frequency synthesizer circuit principles

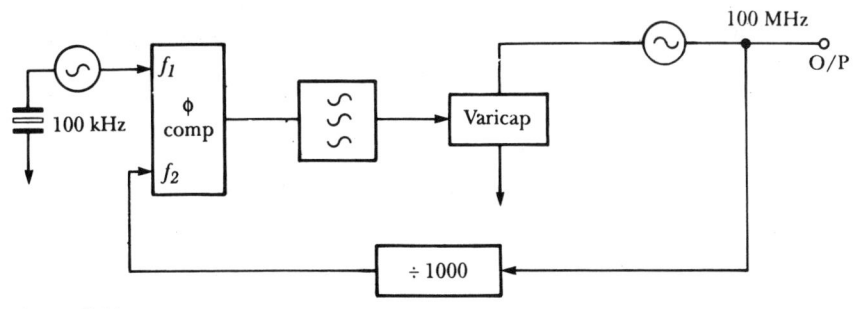

Figure 7.14

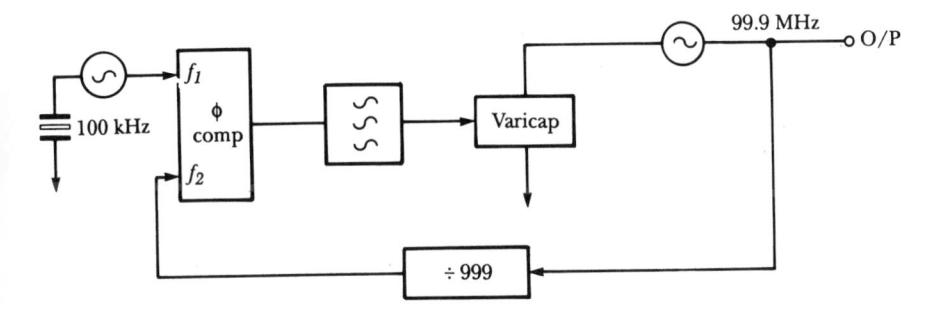

Figure 7.15

The VCO is sitting at 100 MHz and the loop is locked. The division ratio is changed to 999 and the frequency output of the divider will change to 100.1001 kHz. As this is not in phase with the 100 kHz reference, the output of the phase comparator will change, hence the DC correction will change, the varicap will change and the VFO frequency will change until the loop locks again. For the output from the divider to be at 100 kHz, given the division ratio of 999, the VFO frequency will be controlled down to 99.9 MHz and the loop will lock.

An important point to understand is that when the divider was at 1000 the VFO had to be at 100 MHz for the loop to lock. Now the divider is at 999 the VFO must be at 99.9 MHz for the loop to lock. This means that the capacitance of the varicap diode (the device controlling the frequency of oscillation) must have a different value, i.e. a greater capacitance for the lower frequency. But, this capacitance can only be different if the DC voltage controlling it is different and the DC voltage can only be different if the pulses out of the phase comparator are different. This means that although in both cases of lock the frequency to input 1 of the phase comparator is 100 kHz, the phase must be different to give the different width of pulses out. The key point to grasp is that each time the loop is in lock it will stabilize with a new phase difference between the two inputs of the phase comparator.

Mobile Radio Servicing Handbook

To return to the original point, it has been seen that by changing the division ratio in the digital divider the VFO frequency has been changed. This gives an easy method of frequency output selection. Note that for each change of the divider the VFO frequency will change by an amount equal to the reference frequency.

The fact that there is a phase difference between the inputs of the Comparator when in lock means pulses will remain on its output to produce the correction voltage. This is undesirable as we wish to generate a very clean low noise output from the frequency synthesiser, and hence must take precautions to filter out these pulses from both inside and outside the loop. The system used in practice overcomes this problem by ensuring that when the loop achieves lock there is again no phase difference between the two inputs of the comparator. This is done by replacing the simple loop filter with an integrator. The pulses produced when the loop first moves out of lock now 'pump up' the integrator to produce the desired d.c. correction voltage. The corrected state is achieved when the correct unchanging d.c. voltage is produced by the integrator, i.e. there are no more pulses going into the integrator, i.e the inputs to the comparator are in phase. The only pulses required now into the intergrator are to make up its loss or leakage which, with the correct choice of components will be very small. These pulses will, in fact, occur as very thin spikes containing very little energy and hence not causing a significant problem.

The phase comparator will need to have an output not only when small changes happen, but also when large loop changes occur, i.e. the comparator inputs will be at different frequencies. For this reason the phase/frequency comparator is fairly complex, having both phase and frequency analysis capability and also a bi-directional output voltage to correct for change in both directions.

Dual modulus prescaling

This section will consider the requirements of the digital divider in the phase locked loop and the need for programming.

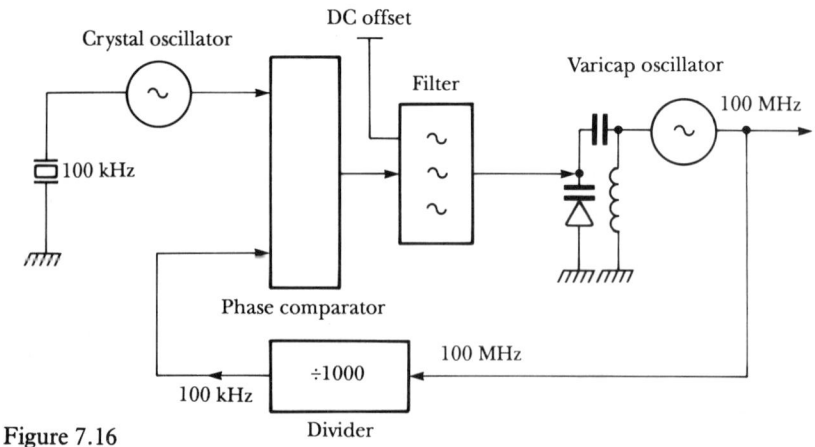

Figure 7.16

Frequency synthesizer circuit principles

To summarize the loop action, for the loop to be in lock the frequency at the input to the phase comparator must be the same and held at some constant phase difference to yield this frequency from the VFO when it is divided by the digital divider. When the number in the divider is changed the phase difference will be altered to give a new VFO frequency to yield the correct frequency (the reference frequency) fed back to the phase comparator.

A reference frequency of 100 kHz has been used in the examples so far, and it has been shown that for a change in the divider ratio of one, the VFO will change by an amount equal to this reference frequency. In a typical PMR system the channel spacing (and hence the frequency steps) is not 100 kHz but 25 or 12.5 or even 6.25 kHz. In order to make identical hardware suitable for all these channel spacings, the equipment manufacturer makes the phase locked loop step in 6.25 kHz steps. This step size will be achieved by a change in the digital divider ratio of one each time. If larger channel spacings which are a multiple of 6.25 kHz are required, then the divider ratio will need to be changed by two for 12.5 kHz and four for 25 kHz for each channel step.

From the above it is seen that the PMR synthesizer needs a reference frequency of 6.25 kHz. Earlier we showed that oscillators of low frequency need large expensive crystals and 6.25 kHz is a very low crystal frequency. This problem can be overcome by making the reference frequency oscillator a much higher value and dividing it down to 6.25 kHz before it is presented to the phase comparator. A crystal of typically 8 to 15 MHz is of optimum size and cost and as the reference frequency divider is a digital component it is easier to divide by a binary number. Modern CMOS circuits are quite capable of working at these frequencies in high division ratios, so a typical reference frequency crystal might be 12.8 MHz with a reference frequency divider of 2048 (2^{11}).

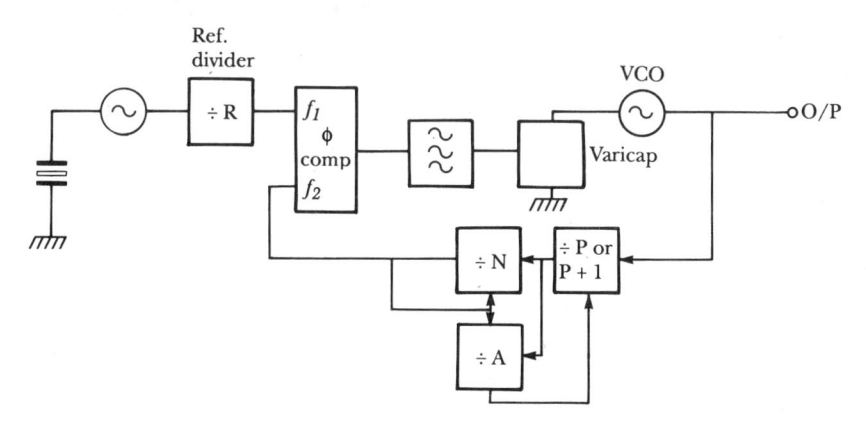

Figure 7.17

There is a further advantage of this division technique. It has been seen that oscillators using crystals have a low noise content in the output frequency.

Mobile Radio Servicing Handbook

There is, however, some degree of noise in the output as the crystal Q is not infinite and the active device (fet or transistor) in the oscillator contributes noise as well. The effect of a divider following the oscillator is to also 'divide down' this noise content and so with a division of 2048 a considerable clean up effect of the reference frequency is obtained.

The other divider to be considered is the frequency select divider in the feedback path between the VFO and the phase comparator. To specify its required parameters, we need to consider the PMR system hardware. In order to fulfill the philosophy of common hardware for all bands, a frequency synthesizer is needed which will give a wide range of frequencies. A frequency of 160 MHz to 500 MHz would enable the equipment to cover several bands. If the reference frequency is 6.25 kHz, a division ratio of 160 MHz/6.25 kHz to 500 MHz/6.25 kHz is needed, i.e. 25 600 to 80 000. Furthermore, the input frequency to the divider is the VFO frequency, i.e. 160 MHz to 500 MHz.

To summarize the divider requirements:

Division ratio = 25 600 to 80 000 in steps of one
(called Nt)
Operating speed = at least 500 MHz

Quite a divider!

One technology which is widely available and is in production at sufficiently low cost and will operate at this speed, is emitter coupled logic or ECL. The preferred large scale integration (LSI) low power technology is CMOS, but this has a maximum operating speed of about 30 MHz. It would seem as though the answer is to prescale the VFO frequency with an ECL divider to bring it below 30 MHz and then use a CMOS programmable divider for the channel or frequency selection. A suitable division ratio would be 500 MHz/25 MHz, ie 20.

Phase comparator ← [÷ n] — [÷ 20] ← VCO

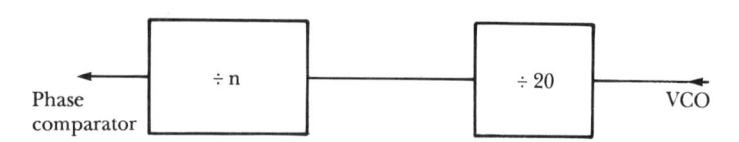

Figure 7.18

There is however one basic flaw in the solution. It is required to divide in steps of one but, with this system, for every change of one in the CMOS divider, the total division ratio changes by twenty. A radio system which can use only every twentieth channel is severely limited in its usefulness.

Note, it may be thought that the reference frequency could be one twentieth of 6.25 kHz, i.e. 312.5 Hz. The problem is that the loop filter would have a very long time constant to remove these pulses from the DC correction voltage and would offer little stability to the VFO and have extremely long lock times.

Frequency synthesizer circuit principles

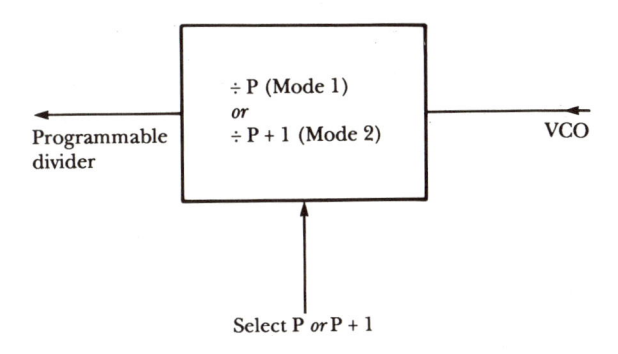

Figure 7.19

There is however an answer to this dilemma. The method used is known as dual modulus prescaling. As ECL and CMOS technology cannot be mixed on one chip, a separate IC is used for the prescaler, this is usually an eight pin dual-in-line to save board area. The method used calls for this prescaler not to divide by one fixed number but to be able to divide by two consecutive numbers, e.g. 10 and 11 or 40 and 41 or 80 and 81.

The prescaler has an input for the high frequency signal from the VFO and a control input that instructs it to divide by either one fixed number (this is called P) or by the second fixed number P + 1. This control input is at a high DC level for one condition or taken to 0V DC for the other condition. In logic terminology it is said to be at '1' or '0' or a 'high' or 'low'. So for example the prescaler may divide by forty when the control input is taken high, and change to divide by forty-one when the control input is taken low. There is also the output of the prescaler. A prescaler of this type is called a dual modulus prescaler.

If the prescaler is used in conjunction with two other dividers, a total divider can be constructed which will fulfill the initial requirements of Nt.

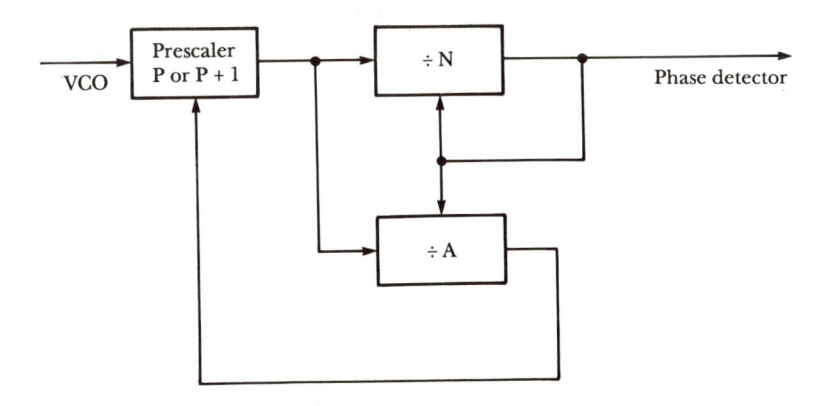

Figure 7.20 *Dual modulus prescaling with A and N counters*

It should be noted here that, although we have discussed dividers performing division, these elements can be considered as counters performing division, e.g. a counter set to count up to sixty-four will give an output at every sixty-fourth input pulse, i.e. it is dividing its input by sixty four. Similarly a counter which is set to contain a value of sixty-four if made to count down instead of up will give an output when the count value reaches zero. It is dividing the input by sixty-four.

So to return to the dual modulus prescaler block, a useful programmable divider (or counter) for the requirement of Nt is made up of an ECL, two modulus dividers and two CMOS counters. A dual modulus prescaler is a counter which can be set to divide by P or by (P + 1). If such a device is placed in a counter configuration with two further dividers, a useful programmable counter for Nt is constructed.

Let the two additional counters be called the A and N counters. We can draw boxes to represent the counter contents and show how the integer steps are achieved.

Step 1 Load the N and A counters. The A counter contents must be less than the N counter contents.

Step 2 Set the prescaler to P + 1. For each P + 1 count that the prescaler performs, decrement both the A and N counters by 1, until the content of the A counter is zero.

The total count so far is

$$Nt = A(P + 1)$$

Step 3 The difference between the N and A original numbers now remains in the N counter, which is (N − A). We now change the prescaler divide number to P and for each time the prescaler performs a P count we decrement the N counter by 1. This process continues until the content of the N counter is zero. The total is

$$Nt = A(P + 1) + (N - A)P$$
$$= AP + A + NP - AP$$
$$= NP + A$$

Some numbers

Let $P = 64$ so $P + 1 = 65$
Let $N = 10$ and $A = 7$
$Nt = NP + A = 10 \times 64 + 7 = 647$

Note that A is free to assume, 0, 1, 2 etc. so a system of programming a counter in integer steps using DMP is achieved.

Frequency synthesizer circuit principles

	Value in N counter	Value in A counter	Set prescaler to:
Step 1 Load N and A	10	7	P + 1 (65)
Step 2a	9	6	
Step 2b	8	5	
Step 2c	7	4	
Step 2d	6	3	
Step 2e	5	2	
Step 2f	4	1	
Step 2g	3	0	

The total so far is equal to A (P + 1), i.e. = 7 (65) = 455

	Value in N counter	Value in A counter	Set prescaler to:
Step 3	3	0	P (64)
Step 3a	2	0	
Step 3b	1	0	
Step 3c	0	0	

This part of the count is equal to (N − A) P, i.e. (10 − 7) 64 = 3 × 64 = 192
Therefore total count (Nt) = 455 + 192 = 647

From Step 2 Nt = A (P + 1)
From Step 3 Nt = A (P + 1) + (N − A) × P
So Nt = AP + A + NP − AP = NP + A

Using the above numbers, Nt = NP + A = (10 × 64) + 7 = 647

It is sensible only to give A the range 0 to P−1, because if A becomes as large as P, the same result is achieved by adding 1 to the N value.

If P = 64, A has the range 0 to 63 which calls for a 6-bit counter, 2^0 to 2^5. We can call this A0 to A5 as shorthand.

If Nt maximum is 40 000 for a VCO frequency of 500 MHz then N must be of the order of 40 000/64 = 625 which calls for a 10-bit counter, 2^0 to 2^9. We can call this N0 to N9 as shorthand.

Example

VCO frequency = 427.1 MHz
Reference frequency = 12.5 kHz

Mobile Radio Servicing Handbook

Calculate Nt, and the numbers which must be programmed into the A and N counters, assume prescale divide numbers are 64/65.

1 Calculate Nt = 427.1 MHz/12.5 kHz = 34168
2 Divide Nt by P = 34168/64 = 533.875. So N = 533
3 For A, multiply fraction by P = 0.875 × 64 = 56
4 Check Nt = NP + A = 533 × 64 + 56 = 34168

In the explanation we have referred to the need to load counter A and counter N with numbers prior to starting the count down process. This means these counters have inputs through which this starting number can be loaded. These inputs are called the programming inputs and we refer to this process of loading a number in the counter as 'programming the counter'.

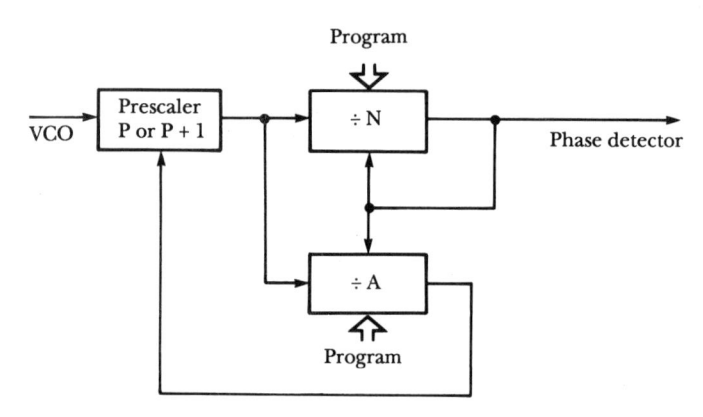

Figure 7.21 *Dual modulus prescaling with A and N counters*

The counters count in a binary sequence. This means that the programme inputs are arranged in a binary weighted sequence. If we assign decimal values to the binary numbering system we can see how this works:

Binary number	2^6	2^5	2^4	2^3	2^2	2^1	2^0
Decimal number	64	32	16	8	4	2	1

So the decimal value 8 can be represented by 2^3 in binary notation. Decimal 64 is binary 2^6. Decimal 12 is binary 2^2 plus binary 2^3. Each of the binary values given above is available as an individual programming input to the counter. We indicate to the counter if a particular binary value is to be used by addressing it with a '1' on the appropriate programming input. Values which are not used are addressed with '0' on the appropriate program input, e.g.:

If decimal 8 is to be loaded the input would be programmed as follows:

Value	2^6	2^5	2^4	2^3	2^2	2^1	2^0
Decimal value	64	32	16	8	4	2	1
Program input	0	0	0	1	0	0	0

Frequency synthesizer circuit principles

If decimal 64 is to be loaded the input would be programmed as follows:

Value	2^6	2^5	2^4	2^3	2^2	2^1	2^0
Decimal value	64	32	16	8	4	2	1
Program input	1	0	0	0	0	0	0

If decimal 12 is to be loaded the input would be programmed as follows:

Value	2^6	2^5	2^4	2^3	2^2	2^1	2^0
Decimal value	64	32	16	8	4	2	1
Program input	0	0	0	1	1	0	0

In practice as the division ratios to be progammed are large, there may be 10 or 12 program address lines for the A and N counters.

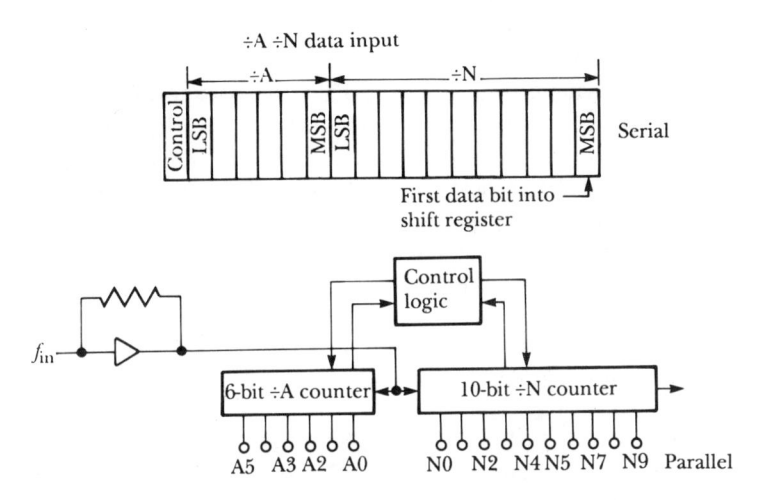

Figure 7.22

Two main methods are used to programme synthesizers, serial and parallel programming. With serial programming the 16-bit programming word is clocked into a shift register whose parallel outputs directly program the A and N counters. With parallel programming the 16-bit words are held in a ROM in the form of two 8-bit or one 16-bit EPROM or on a diode matrix board. Note that a 16-bit data word is necessary for each frequency, so that two words are required for each channel on which the transceiver operates. Some synthesized radios have a fixed frequency difference between receive and transmit and may therefore require only one data word per channel; they generate the companion word on chip.

Mobile Radio Servicing Handbook
Practical considerations

The requirements and actions of a phase lock loop have been fully considered. The change in frequency of the VFO in the frequency synthesizer is made by changing the capacitance of the varicap.

The method of changing channel is by changing Nt. As the reference frequency remains constant the VCO moves by virtue of an error voltage appearing at the output of the phase comparator. This voltage is taken through the loop filter to the VCO. The time taken by the synthesizer to change channel must be less than about 50 msec, otherwise information will be lost on some signalling systems. Changing channel by suddenly changing the value of Nt is one form of disturbance.

In Chapter 4 the different forms of modulation were examined and it was concluded that FM is frequently used for PMR. To frequency modulate the transmitter, a method of changing the carrier frequency about the nominal channel centre in sympathy with the modulation waveform is needed. A customary method of doing this is to add a second varicap diode in the oscillator tuned circuit. As the capacitance of this is varied by the modulating waveform, the VCO frequency will vary as well. However it must be remembered that the whole purpose of the phase locked loop is to prevent the VCO from changing frequency due to an external influence.

The correcting path is through the loop filter so that if the change is faster than the loop can follow the correction action will be lost in the filter. This means in designing the loop the time constant of the loop filter must be made longer than the lowest modulating frequency so that it is not removed or 'wiped off' by the loop.

There is a conflict here, to have the loop filter slow enough not to wipe off the modulation at say CTCSS frequencies of 60 Hz would mean that it would take about 1.5 seconds to change channel which is clearly unacceptable. Thus a way of speeding up the loop response must be found when changing channel. A simple speed up circuit is shown and works by taking advantage of the fact that the loop goes out of lock when changing channel but not when being modulated by speech. When out of lock, a very large error signal appears from the phase detector, and if this error voltage exceeds about 0.6 V in either polarity, the appropriate transistor turns on and by-passes most of the series resistance in the loop filter. This greatly speeds up the loop, and when the loop is nearly in lock the error voltage falls below 0.6 V and the filter assumes its normal much narrower bandwidth. The error voltages at this point caused by the modulation are insufficient to turn either transistor on.

When changing from receive to transmit the VCO is changed from being used as the receiver LO to generating the transmit frequency. This means the loop will have to swing through a frequency equal to the IF, typically 10.7 MHz or 21.4 MHz. In order to save the loop lock time which this would require, some radio systems use two VFOs in the loop.

Frequency synthesizer circuit principles

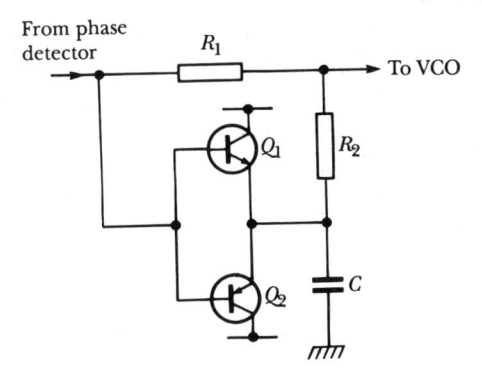

Figure 7.23 *Simple speed up circuit*

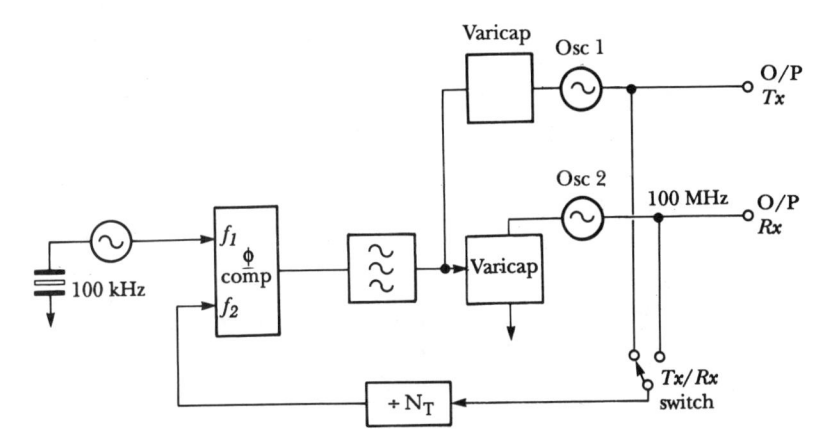

Figure 7.24

Both have their outputs connected to the prescaler input and both have their varicap control voltage lines connected to the loop filter output. The first oscillator has its inductance chosen so that when the loop correction voltage is at mid point, the oscillator is running at the receiver LO frequency. The second oscillator's inductance is selected so that at mid point DC loop voltage the oscillator is running at the transmit frequency. This means that when switching from receive to transmit, the DC correction has very little voltage to travel through to achieve lock as the appropriate oscillator is already running at approximately the right frequency. This speeds up the receive/transmit lock time considerably.

8 Principles of RF measurement

Test and measurement objective

There are many receiver, transmitter and system performance measurements which can be made. Each test or measurement is suited to examining a particular parameter in the equipment and to achieving a certain objective. The objectives with which the test and service engineer will most commonly be concerned are:

- Pre-delivery checks
- Calibration
- Acceptance testing
- Installation and commissioning
- Post-installation checks
- In-service fault finding

Some tests are suitable for more than one requirement. Other tests are specific to a particular requirement. The engineer should regard the various tests available to him as tools and his job is to pick the best tool for the job and then to use it correctly. The correct use of a test and the interpretation of the results obtained can only be achieved through an understanding of the objectives of the test and a knowledge of what is happening under test conditions.

RF fundamentals

Noise

The prime parameter used to evaluate receiver performance is sensitivity. The maximum usable sensitivity (MUS) is defined as the minimum signal at the receiver input which will produce an audio output signal with a 12 dB SINAD ratio (see later in this chapter for further information on SINAD) under defined modulation conditions. The term SINAD refers to the ratio of measured signal and noise distortion to noise and distortion, expressed in dB. In a well designed, correctly working receiver the distortion products produced will be far less significant than the noise part of the equation. In such a receiver the noise figure is the sum of the RF amplifier device noise plus all the losses before it (in dB). The MUS also comprehends noise entering the antenna system with the wanted signal.

The total noise under consideration can be galactic, atmospheric, man-made or noise due to losses in the antenna transmission line system or thermal noise in the receiver. Below approximately 20 MHz atmospheric noise predominates. Between 20 MHz and 100 MHz galactic noise predominates. As equipment up to these frequencies tends to use larger, more sensitive antennas, this natural received noise is the dominant factor in the total receiving system and extreme low noise front end design has little value.

Atmospheric noise is often swamped by man-made noise in cities, but in rural locations is a more significant factor. It is caused primarily by natural static conditions, e.g. lightning, and so is both weather and geographically dependent.

Galactic noise is noise which has its origin beyond the earth or its atmosphere. It is caused, in part, by solar activity and so is affected by sun spot action, and partly by the various RF noise sources existing in local galaxies.

At frequencies greater than 100 MHz the galactic and natural noise falls away. This means the amount of noise being received with the VHF/UHF signal is reduced. As VHF/UHF equipments use smaller, less efficient antennas, the noise *generated* within the antenna and receiver predominates. At higher VHF frequencies and UHF, antenna noise falls off and receiver performance becomes more important.

The noise generated within the receiver system is due to two main factors:

1 Losses

- *Line losses* This includes all losses in the receiving system up to the point of first gain in the receiver. When a signal passes through antennas, couplers, connectors, transmission lines etc., it will suffer attenuation. This effectively reduces the signal-to-noise ratio and so increases the receiver noise factor.
- *Loss resistance in a warm line* The existence of attenuation in itself implies loss resistance. A warm transmission line will experience thermal agitation and thermal noise. As the temperature increases, the random motion of electrons in the conductor increases. The minute voltages created by the electron motion constitute the thermal noise.

2 Thermal

The thermal noise effects in a receiving system are the prime limiting factors in the RF amplifiers, mixers and first IF stages when handling low level signals.

Thermal noise is created by free electron movement, i.e. it is a direct result of current flowing. This random electron movement gives potentials across the circuit conductors. As this effect was demonstrated by J. Johnson, it is often referred to as Johnson noise. Noise is a random effect and the total 'noise power' generated in a circuit is a direct function of its bandwith. The only way to minimize this effect in active circuits is by careful semiconductor device design and well controlled circuit parameters. Where extreme low noise amplification is required, i.e. satellite systems, it is not uncommon to run the amplifier immersed in liquid nitrogen for extreme low temperatures.

Intermodulation distortion

An amplifier should ideally act as a linear device, i.e. the output voltage Vo should be proportional to the input voltage.

$$Vo = AV_{IN}$$

In practice, output characteristic is not linear and the curvature can be expressed as

$$Vo = AV_{IN} + BV_{IN}^2 + CV_{IN}^3 + \ldots$$

The terms BV_{IN}^2, CV_{IN}^3 . . . give rise to second third, etc, order distortion.

If V_{IN} is doubled, the second order terms increase 4 times (2^2) and the third order 8 times (2^3), ie as the input is increased the ratio of distortion to wanted signal increases.

If two equal signals are applied, the BV_{IN}^2 term will produce distortion of $2f_1$, $2f_2$, $f_1 + f_2$ and $f_1 - f_2$, and the BV_{IN}^3 term will produce distortion of $3f_1$, $3f_2$, $f_1 + 2f_2$, $f_1 - 2f_2$, $2f_1 + f_2$, $2f_1 - 2f_2$ etc.

Hence, when a complex wave is applied to a non-linear device, a noise spectrum will be produced. This is referred to as 'intermodulation distortion' (IMD). As a rule of thumb, a good receiver should have IMD products at or below the receiver noise level.

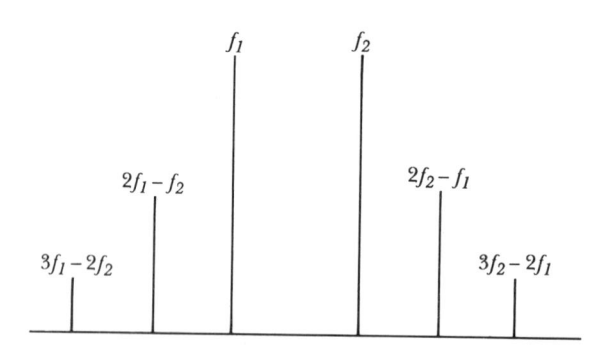

Figure 8.1

We can compare the IMD attenuation performance of two receivers by comparing their third order intercept points, the point at which the two-tone third order response equals the two-tone input.

If we cannot ascertain the third order intercept point, we can judge the linearity of the amplifier/mixer by looking for the 1 dB compression point.

As RF input is increased into the receiver, we should see a parallel linear increase in IF output. If the IF output begins to increase at a lower rate, we know the amplifier or mixer stage is going into compression. Deviation from the linear curve by 1 dB is the 1 dB compression point. For example, if we have a generator input of -50 dBm and an IF output of -30 dBm, we know our gain is 20 dBm. If we increase our generator input in 10 dBm steps until we arrive at -20 dBm, we would expect 0 dBm from the IF output. If the IF output is only -1 dBm, then we know we have reached the 1 dB compression point. The third order intercept point is generally 10–15 dB higher than the 1 dB compression point.

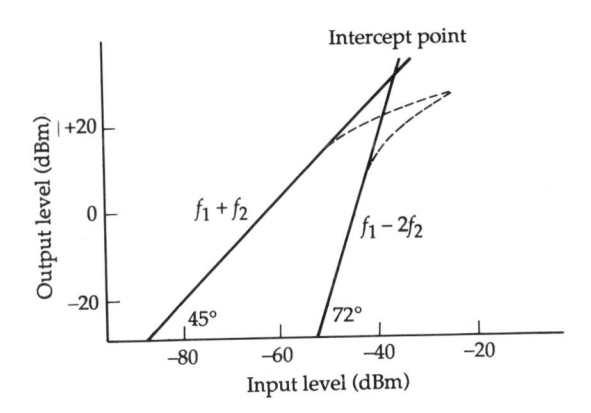

Figure 8.2

Cross modulation

Cross modulation occurs when the modulation from an unwanted signal is transferred to the wanted signal. The effect is proportional to the square of the unwanted signal amplitude. A performance improvement can therefore often be made by attenuating the signal input at the expense of noise performance due to the inserted loss.

Cross modulation can be measured using two signal generators. One with 30 per cent AM, the other with a CW signal. The output of the AM generator is increased until 1 per cent modulation appears on the second generated CW signal measured on a spectrum analyser.

Note, when measuring cross modulation, it is important to remember that many signal generators create their own additional sidebands, requiring additional crystal filtering. It is also essential to ensure that the amplitude modulated signal generator has negligible spurious FM.

Also, when two or more signal generators are used together, they should be connected to the UUT (unit under test) via a hybrid combiner, a unit with three or more ports giving 6 dB of attenuation between the input and output ports and 40 dB of attenuation between the input ports. A hybrid combiner prevents one signal generator frequency or phase modulating a parallel signal generator, and maintains a 50 ohm impedence level throughout the system.

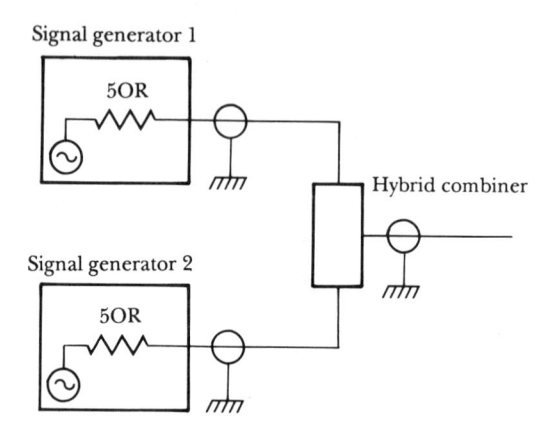

Figure 8.3

Gain compression

Gain compression occurs when an adjacent strong signal causes an apparent decrease in receiver gain. The input voltage from the unwanted signal forces the amplifier/mixer to lose gain response.

This can be measured by using two signal generators a defined frequency spacing apart. The second generator is adjusted until the wanted signal from the first generator becomes depressed by a defined amount, usually 3 dB.

Dynamic range

The dynamic range of a receiver can be defined with the upper limit being the input level required to produce third order intermodulation distortion products equal to the receiver noise level and a lower limit being the noise floor figure of the receiver.

Time and frequency analysis

In any comparison of circuit performance, we have the choice of using three

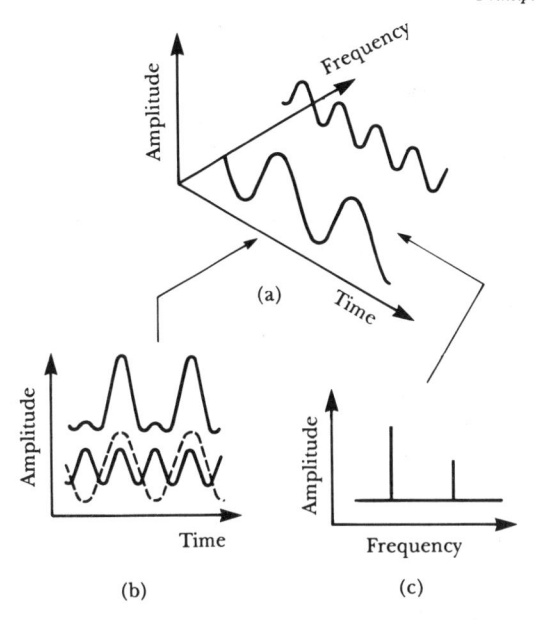

Figure 8.4

dimensional co-ordinates – time or frequency against amplitude. An oscilloscope displays signals in the time domain showing amplitude versus time. A spectrum analyser displays signals in the frequency domain showing amplitude versus frequency. It is optimized to characterize the signal by displaying the relative amplitudes of its frequency components, but is unable to display phasing information.

The difference between the time domain and frequency domain can be shown by the example of a tuning fork at audio frequencies.

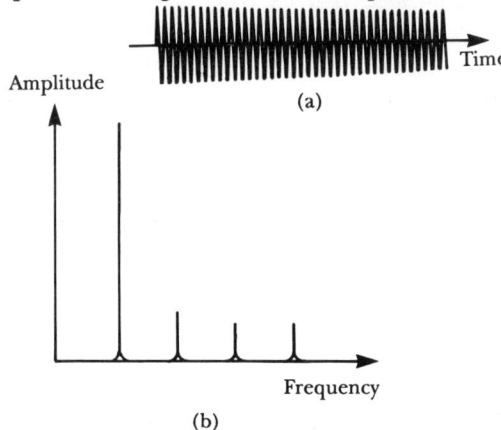

Figure 8.5 (a) *Time domain view of the sound from a tuning fork, (b) frequency domain view of the sound from a tuning fork*

The time domain view of the tuning fork (Figure 8.5a) shows the amplitude decreasing over a period. The frequency remains the same. The frequency domain view (Figure 8.5b) shows the frequencies individually given out by the tuning fork. Although from Figure 8.5a we might conclude that the waveform is a perfect sinusoid, Figure 8.5b shows that it is, primarily a single frequency with distortions produced by small frequency multiples.

Over one hundred years ago, Fourier demonstrated that any waveform can be generated by the addition of a number of harmonically related sine waves (Figure 8.6).

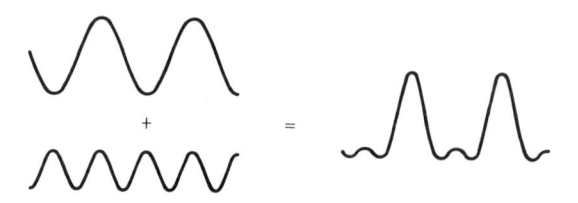

Figure 8.6

Working back from this, we can see that frequency analysis allows for the visual separation of the sine waves that make up a complex waveform.

If we look at a sine wave along the frequency axis, each sine wave frequency appears as a vertical line with height representing amplitude and its position representing frequency. We use frequency analysis to represent the spectrum of the signal – hence the term spectrum analysis.

The application of frequency analysis is shown, for example, in the measurement of distortion in an audio oscillator where a small sine wave distortion component has to be detected in amongst larger signals.

The small signal is hidden in a time domain measurement (oscilloscope) within a single sine wave, but shows up in the frequency domain measurement (spectrum analyser) as a separate subsidiary peak.

The human ear and brain works as an audio spectrum analyser, dividing the audio spectrum into narrow bands and determining the power present in each band.

Frequency response

In any electrical network, gain, loss and phase shift is a function of frequency. In a linear network, frequency response is independent of the input amplitude. The output of any network can be determined by its frequency response which can be low pass, high pass, band pass or a combination of all three. Multiplying the spectrum of the input signal by the response of the network will define the characteristics of the output signal. These are steady state responses, but a network will also have a transient response to an input signal.

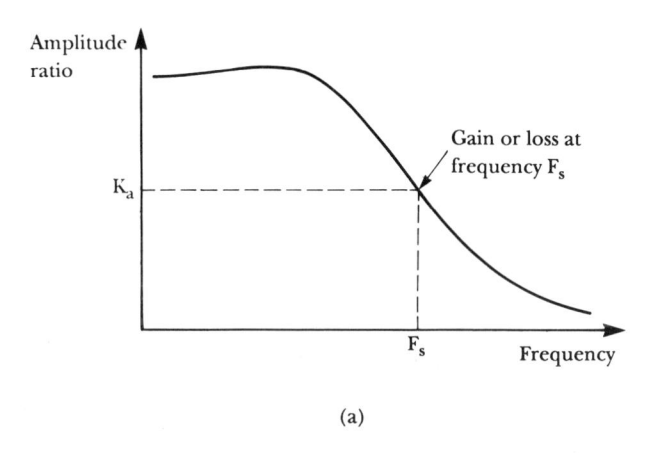

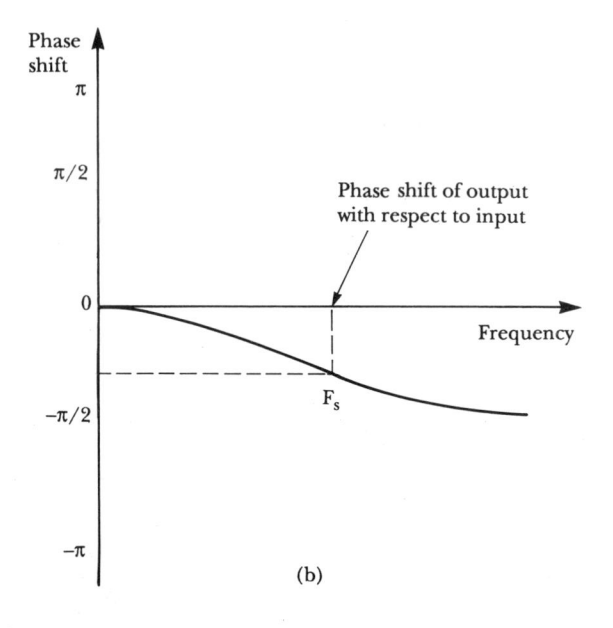

Figure 8.7

Mobile Radio Servicing Handbook

The response characteristic of a tuned circuit is defined as its Q where Q equals the centre frequency of resonance divided by the frequency width of the −3dB points. A high Q filter will have a slow decay transient response. A low Q filter will have a fast decay transient response.

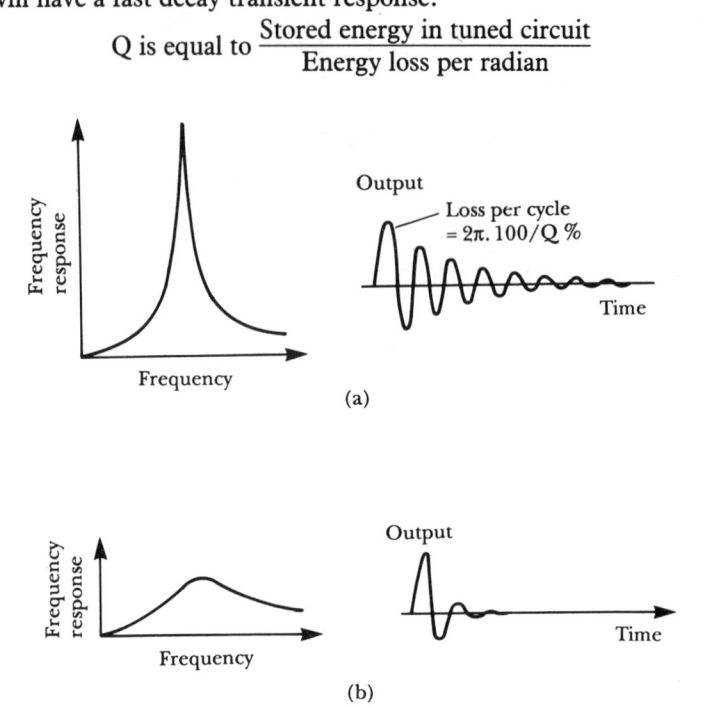

$$Q \text{ is equal to } \frac{\text{Stored energy in tuned circuit}}{\text{Energy loss per radian}}$$

(a)

Loss per cycle = $2\pi \cdot 100/Q$ %

(b)

Figure 8.8 *(a) Lightly damped (high Q) filter, (b) heavily damped (low Q) filter*

In principle, signals that are broad in one domain are narrow in the other. Narrow selective filters, for example, will have long response times. The design of a spectrum analyser is in itself dependent on these filter characteristics.

A parallel filter spectrum analyser uses a large number of filters which provide for fast measurement of the signal, but limited resolution. The resolution is dependent on the number of filters used, the limitation here is normally cost!

A swept spectrum analyser, as most commonly used in RF and microwave spectrum analysis, uses one filter and sweeps across the required spectrum band width. This provides for high resolution (closely spaced spectral line analysis), but adds a time component, i.e. it has to assume that the signal is unchanged during the sweep. This may mean the analyser misses transient events. The sweep speed in itself is limited by the response time of the filter.

If the sweep speed is too high, the signal will not remain within the pass band

Principles of RF measurement

for sufficient time to build up the response. As a rough guide, the time for which the signal should be within the pass band should not be less than

$$\frac{1}{\text{bandwidth of filter}}$$

A filter bandwidth of say 10 kHz would require 100µS to allow for build up and decay time. This limits the sweep speed to 10 kHz in 100µS, i.e. 100 MHz/second. The limitation becomes more serious in audio spectrum analysis where bandwidths of only a few cycles are required, which can limit sweep speeds to less than 100 Hz/second. This is a feature that must be appreciated when using a spectrum analyser if very misleading results are to be avoided.

Fast digital signal processing now makes possible the transformation of data from time domain to frequency domain using fast Fourier transform analysis.

Use of decibels as a measurement base

When resolving small signals in the presence of large signals, it is commonly required to represent a distortion component (harmonics etc.) that is as small as 0.1 per cent of the primary signal. We have the problem of how to show this ratio graphically.

A logarithmic scale acts as a compander – compressing large signal amplitudes and expanding small amplitudes. Alexander Graham Bell, needing to express the ear's logarithmic response to power difference, produced the unit of a bel, and the 1/10 bel or decibel as a logarithmic unit.

The bel is a power ratio. Powers P_1 and P_2 have a ratio

$$\log_{10} \frac{P_1}{P_2} \text{ bels}$$

i.e. if $P_1 = P_2$, the ratio is 0 bels
if $P_1 = 10P_2$ the ratio is 1 bel (10 dB)

The decibel is frequently used to express voltage or current ratios. Since power is proportional to (voltage)^2, the ratio of voltage V_1 and V_2 is given by

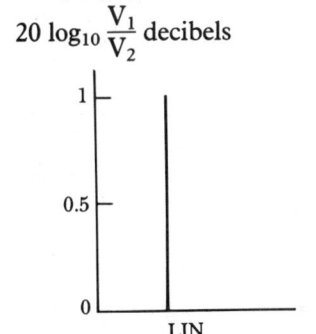

$$20 \log_{10} \frac{V_1}{V_2} \text{ decibels}$$

(a) 0.1% not visible on scale

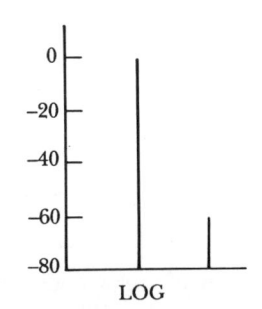

(b) 0.1% shown as 60 dB down on fundamental

Figure 8.9

Our 0.1 per cent distortion component expressed in dB is 60 dB below the fundamental frequency.

In typical RF work, power will usually be specified in dB (decibels) with respect to one milliwatt and designated dBm, i.e. 0 dBm will equal 1 milliwatt in a 50 ohm system.

A typical transmitter output of 2 watts is expressed as +33 dBm. A typical signal arriving from a 50 ohm antenna at the input of a receiver would be 1 microvolt or −107 dBm.

Receiver parameter definitions

As we have seen from the section on noise and active circuit limitations, we can assess the performance of a receiver on the basis of sensitivity (noise related) and selectivity, including the receiver's immunity to inter-modulation distortion, cross-modulation and gain compression (blocking). In addition we can assess the receiver's stability and frequency accuracy. It is worth restating some of the measurement definitions:

Adjacent channel selectivity

The ability of the receiver to resolve a wanted modulation signal from an unwanted adjacent channel modulation signal. Measurement requires two signal generators, one with a modulated 1 kHz tone, two-thirds maximum system deviation and set for 12 dB SINAD. The second generator is modulated with a 400 Hz tone at the same deviation on the adjacent channel. The amplitude ratio of generator 2 compared with generator 1 is measured, when generator 2 is adjusted for a 6 dB SINAD figure. This ratio is the measure of ACS.

Intermodulation spurious attenuation

The ability of the receiver to distinguish wanted signal from combinations of two or more unwanted signals at other frequencies. The measurement method requires three signal generators with the first generator modulated with a 1 kHz tone, two thirds maximum system deviation, and level adjusted to −12 dB SINAD. The second generator unmodulated is tuned to the adjacent channel, and the third generator with 400 Hz modulation, two thirds maximum system deviation on the next adjacent channel. The output of the second and third generators needs to be increased equally until the SINAD degrades to 6 dB. The ratio of the level of generator 2 to generator 1 then gives us the intermodulation spurious attenuation.

Cross modulation

Cross modulation occurs when the modulation from a strong unwanted signal close in frequency modulates the wanted signal. It again occurs due to non-linearity in front end circuits.

Reciprocal mixing

Mixing which can occur (particularly in frequency synthesized sets) where noise sidebands appear on the local oscillator and mix with the incoming RF signal to give unwanted products in the IF passband.

Signal to noise ratio

The amount of signal required to be 10 dB for AM or 12 dB for FM greater than the internally generated noise of the receiver.

AGC range

The measure of change in the audio output against input signal level change.

Desensitization

The effect of a strong signal forcing the front end of the receiver into gain compression, and so reducing the level of small signals.

Selectivity

Selectivity is the frequency bandpass response of the various signal sections in the receiver – RF, IF and AF. This response assists in excluding unwanted frequencies. It can be measured using swept frequency or wideband noise techniques.

Sensitivity

Sensitivity measurements are covered in detail in the later section on SINAD (p.98). One point to note on both SINAD and quieting measurements, is to ensure that a proper impedance match is achieved between generator and receiver. This normally involves using an attenuator pad which smooths out the reflected impedance variations.

Transmitter parameter definitions

Field service transmitter measurement parameters are dealt with in detail in the DTI specification MPT 1372. Summary points from the specification are, however, as follows:

Test equipment

Test equipment is required to be of sufficient measurement accuracy. This is defined as less than 5 ppm variation for RF signal generation and measurement, less than 5 per cent error for modulation and power indication, and less than −75 dBc (dB relative to carrier) for harmonic and spurious measurements. The test equipment must fully cover the frequency power voltage range of the equipment under repair and the spectrum analyser should cover at least the third harmonic of the carrier frequency.

Measurement guidelines are as follows:

RF carrier frequency

It is necessary to make sure that the frequency measured is not a harmonic or spurious signal from the transmitter or from the test equipment. The transmitter should be terminated with a dummy load of suitable power rating, frequency range and impedance, and the measurement should be made without modulation applied. Where the transmitter has an integral antenna, the measurement should be made off air using a short rod antenna on the input of the counter. You need to make sure that the reading is not being affected by other transmitters in the proximity.

RF power

The transmitter again should be terminated with a dummy load and the power measured from the transmitter socket without modulation applied. Either a terminated or through-line power meter should be used – a throughline should have an external load. The ERP has to be calculated taking into account the attenuation of the filters, isolators, combiners, feeder cables, and antenna gain.

Mobile unit measurements are described in MPT 1362. As a rule of thumb, the antenna of a hand portable will have a loss of 6 dB compared to a half-wave dipole.

Angle (FM) modulation

It is considered quite acceptable to check deviation limiting by speaking into the microphone. To find the frequency within the audio passband at which peak deviation occurs requires a variable frequency audio signal generator to sweep the passband.

Permissible frequency deviation can be worked out from the approximate band width formula, where $BN = 2(M + D)$ where BN is the necessary band width in Hertz, M is the maximum modulation frequency in Hertz and D is the deviation in Hertz. For example, a 12.5 kHz channel with 11 kHz of band width would have a maximum modulation frequency of 3 kHz and a deviation of 2.5 kHz.

The speech audio deviation may also be affected by the presence of CTCSS and the CTCSS tone deviation level should be checked. Levels are set out in MPT 1306 and are 400–800 Hz for 25 kHz channel separation, and 200–400 Hz for 12.5 kHz separation. The CTCSS tone frequency should also be checked using a frequency counter connected to the demodulated audio output of the modulation meter.

Amplitude modulation

For practical purposes it is generally considered to be sufficient to ensure that no speech peaks exceed 100 per cent. CTCSS can be found on AM systems and in these cases the maximum modulation depth permitted is 10–20 per cent.

Digital signalling

We need to check the deviation of the signalling tones. This can be done either by measuring each tone in turn or displaying the demodulated audio output on an oscilloscope with a vertical axis calibrated in kHz deviation. For FM, ± 60 kHz peak deviation, ± 20 per cent maximum deviation is permissible. For AM, 60 per cent maximum modulation depth, ± 20 per cent is permissible.

Harmonics and spurious

These measurements would typically require a spectrum analyser with a filter between the sampler and the analyser to filter out the fundamental signals. Off air measurements using a short rod antenna on the analyser input can be misleading because of variable coupling with frequency and positioning differences, but can give at least an indication of the spurious signals occurring in the proximity of the carrier.

Receiver tests

Sensitivity

Maximum usable sensitivity (MUS) is the minimum signal level at the receiver input that will produce an audio frequency input having a 12 dB SINAD ratio.

The sensitivity check will determine correct performance of the antenna connector, any harmonic filters, the transmit/receive RF switching, the pre-RF amplifier bandpass filtering and the RF amplifier itself. This is because in a well designed receiver, the noise figure of the receiver is the sum of the RF amplifier device noise figure, plus all the losses before it (in dB), and the 12 dB SINAD (FM) or 10 dB S/N (AM) sensitivity depends on the noise figure.

The 12 dB SINAD test will be examined in detail as it is a most useful check of receiver performance but is frequently misunderstood in its action.

1 *The SINAD test*

The signal level at which a receiver produces a 12 dB SINAD ratio is referred to as the 12 dB SINAD sensitivity of the receiver. In practice, a 12 dB SINAD signal is a reasonably intelligible and useful signal for speech transmission.

Since a SINAD measurement gives a more meaningful measurement of a receiver's useful sensitivity than is obtained by other methods, it has become the preferred method of specifying and measuring receiver sensitivity in FM receivers used in land mobile and marine services.

The exact method of measuring 12 dB SINAD sensitivity is given in the Electronic Industries Association's Standard RS–204–A, which is quoted here:

> 'A 1000 microvolt test signal from a standard input signal source with standard test modulation shall be connected to the receiver antenna input terminals. A standard output load and a distortion meter incorporating a 1000 Hz band elimination filter shall be connected to the receiver audio terminals. The receiver volume control (low level) shall be adjusted to give rated audio output. The standard input signal level shall be reduced until the SINAD is 12 dB. At this value of signal input, the audio output shall be at least 50 per cent of the rated audio output without readjustment of the volume control. If the audio output is less than 50 per cent of the rated audio output, the input signal level shall be increased until 50 per cent of full rated audio output is obtained, and this value of input signal level shall be used in specifying sensitivity.'

Note, a receiver with more than one volume control shall be adjusted utilizing a control preceding the audio power amplifier.

Standard RS–204–A specifies that the receiver shall be operated into a resistive load equivalent to the load into which the receiver normally operates. It also specifies standard test modulation as being sixty per cent of the peak modulation used.

Since the SINAD definition includes the distortion created by the receiver's audio output stage, a precise measurement of SINAD should be made at the rated audio output. However, in typical equipment with low distortion amplifiers, a reasonably accurate SINAD measurement can be made with the audio output merely set at a comfortable listening level, using the loud-speaker of the receiver as the audio load.

Principles of RF measurement

2 How the SINAD test works

The purpose of a radio receiver system is to intercept a transmitted signal, to process the signal until suitable for demodulation, to demodulate it and to present this demodulated signal to an output loudspeaker. All this processing must be performed without distorting the original signal and adding anything to it, i.e. noise, harmonics etc. The purpose of the SINAD test is to check whether the receiver under test performs the above task correctly. The SINAD test requires that a well defined (i.e. known) signal is applied to the antenna socket of the receiver. This input takes the form of an RF carrier exactly on the selected channel frequency and modulated with a single sinusoidal tone of 1 kHz. The standard SINAD test definition requires a modulation of 60 per cent of maximum system deviation.

If the receiver system is 'Ideal' the 1 KHz tone will be reproduced at the output, i.e. loudspeaker receiver without noise or distortion (Fig 8.11). The SINAD meter measures to what degree the receiver approaches this capability. A description of the SINAD meter system will make clear how the measurement is achieved.

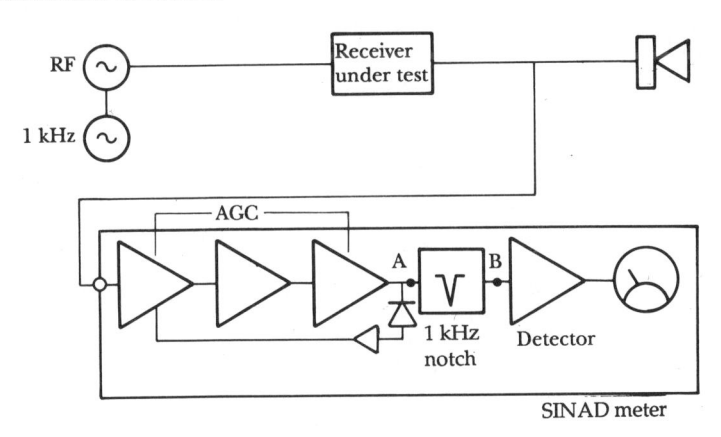

Figure 8.10

The first section of the SINAD meter consists of one or more stages of automatic gain control (AGC) amplifiers. These, with the diode level detector, ensure that the signal at point A is kept at a constant amplitude for a large range of input levels. As the same SINAD meter will be used in the workshop for different makes of radio, some with high impedance speakers and some with low impedance, it is designed to accept input voltage levels from typically 5 mV RMS to 7 V RMS and so not be radio specific.

The waveform at point A will be the 1 kHz sinewave (output at the speaker terminals of the receiver under test) held to a standard amplitude plus any unwanted signals generated in the receiver under test (Fig 8.12). This is now fed to a 1 kHz notch filter. This filter passes the whole of the audio band (300 Hz to >3 kHz) except for a very narrow portion at 1 kHz which it completely blocks.

Mobile Radio Servicing Handbook

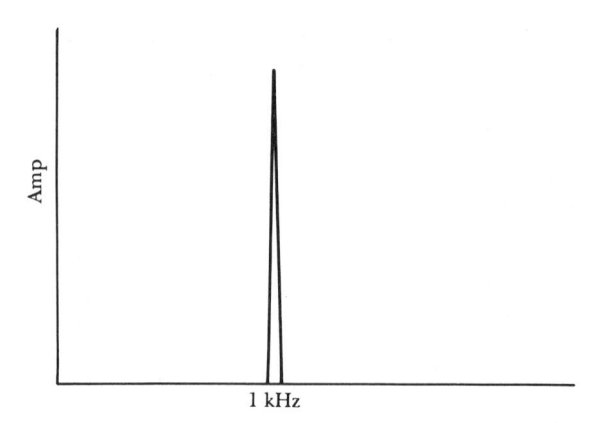

Figure 8.11 *Spectrum of 'ideal' 1 KHz tone*

The purpose of this can be seen by examining the spectrum of the 1 kHz tone from the point at which it modulates the signal generator at the receiver input, up to point B in the SINAD meter.

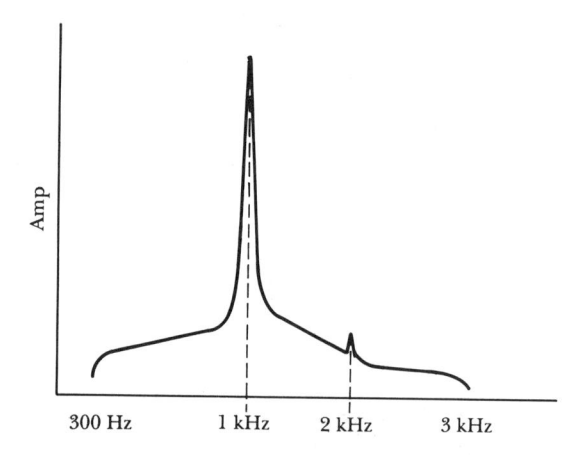

Figure 8.12

Figure 8.11 shows the pure 1 kHz tone modulated on the signal generator at the input of the receiver. Figure 8.12 shows the output at the speaker terminals of the receiver. It is seen that the receiver has added to it a noise level and a very small level of second harmonic.

Figure 8.13 shows the signal at point A in the SINAD meter. As the AGC detector is a peak detector, it will always hold the peak of the signal at a constant amplitude. If the injection level of the signal generator at the receiver antenna is reduced, the noise relative to the signal level will increase (the

Principles of RF measurement

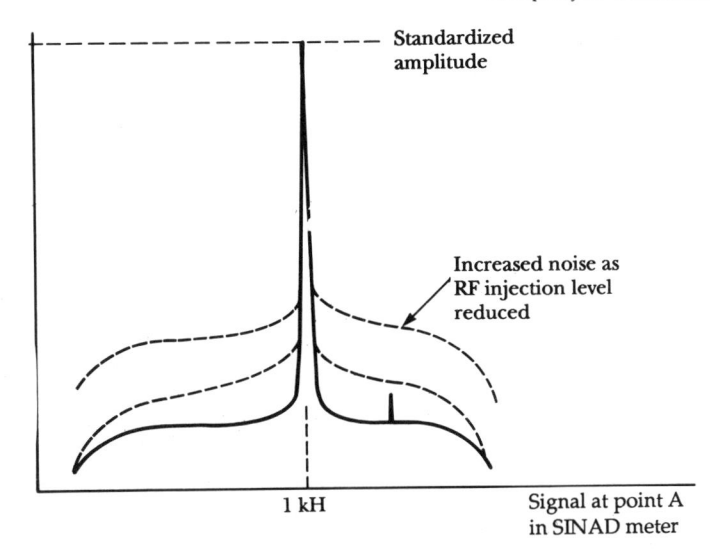

Figure 8.13

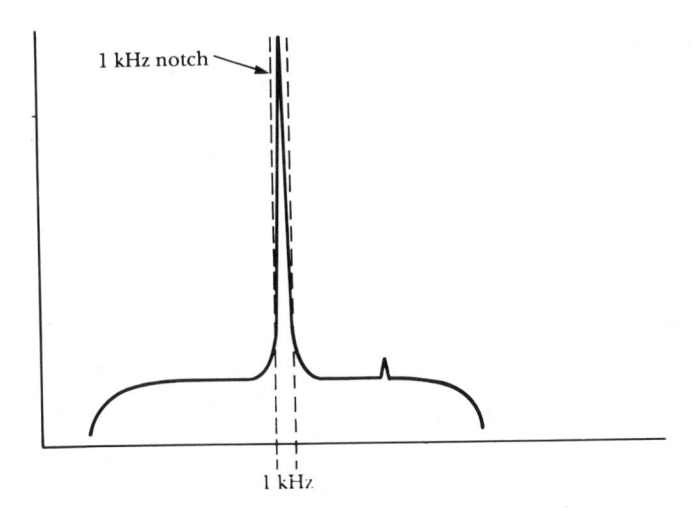

Figure 8.14

squelch should be open) and this is seen in Figure 8.13 as a worsening signal to noise ratio.

Figure 8.14 shows where the response of the filter will occur on the signal presented to its input, i.e. point A.

Figure 8.15 shows the signal at the output of the notch filter, i.e. point B. This signal consists of just those products (noise + distortion) that the receiver has added to its original input. The level of this signal will depend directly on the injection level of the signal generator at the receiver input.

Mobile Radio Servicing Handbook

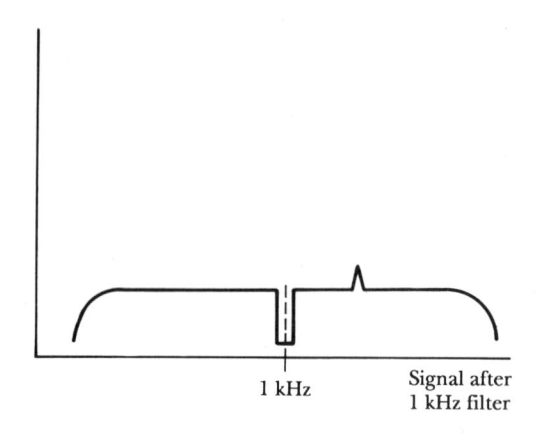

1 kHz   Signal after
          1 kHz filter

Figure 8.15

The signal now passes to a detector – ideally an RMS detector but usually a peak detector. The output of this detector drives a meter. As the level at the output of the AGC section (point A) is set in calibration the meter can be scaled as a ratio between this peak signal level and noise/distortion signal and noise/distortion signal presented to the input of the detector (point B).

3 Measuring receiver sensitivity by the SINAD method

Assume a typical FM communications type receiver for use on channels with 5 kHz peak deviation.

A signal generator is connected to the antenna input of the receiver to apply a signal on the receiver channel, modulated with a 1 kHz tone, at 60% of system peak modulation deviation.

The signal generator level is adjusted to a high output – the SINAD meter should move to the left of its scale (high value of dB). Decreasing the signal amplitude to zero will cause the SINAD meter to move to the right of the scale (low value of dB). The signal generator level is then adjusted until the SINAD meter shows 12 dB. Reading the microvolts or dBm output from the signal generator gives the '12 dB SINAD sensitivity' of the receiver.

Whilst making the measurements, the receiver volume should be adjusted to the receiver's rated audio output to verify that the receiver meets the manufacturer's published specifications. This probably will make it necessary to replace the loudspeaker temporarily with a resistor of the same impedance value.

If one is making a bench sensitivity check, the loudspeaker can be run at a comfortable level.

4 About SINAD accuracy

For a precise determination of the rate implied by the SINAD definition, the measurement circuits of the distortion meter should measure the RMS values of the composite signal, noise and distortion waves. However, almost all commercially available distortion meters are based upon average measuring, but RMS calibrated, metering cicuits. At the low distortion and noise percentages involved in the typical 12 dB SINAD measurements, the error created by the use of average metering circuits instead of RMS metering circuits is negligible.

The width of the null in commercial distortion meters varies considerably from one model to another. Whilst this will not cause any discrepancy in simple distortion measurements, the width of the null will affect readings on noise measurements. Therefore, perfect correlation between SINAD indications may not be obtained between different model distortion meters, although they agree perfectly on ordinary distortion measurements.

Although different model distortion meters may give slightly different SINAD readings on the same composite signal, the SINAD method of measuring receiver sensitivity is remarkably precise. This is because the 12 dB SINAD performance of a typical FM receiver falls in a place on the FM improvement curve where a small percentage change in incoming signal will create a large change in SINAD reading. Thus, distortion meters differing by two or three dB in their SINAD reading will result in 12 dB SINAD sensitivity measurements which correlate to better than 1 dB. In other words a 12 dB SINAD sensitivity measurement made with the SINAD meter will correlate within 1 dB to the sensitivity measurement obtained by the use of the most popular distortion meters.

5 *About the meter flicker*

The flickering of the meter pointer is caused by the statistical nature of the noise in the receiver output. Since this flickering is a basic fact of nature, the only way to reduce it (and still make a true SINAD measurement) would be to slow down the meter response time. The response is, in fact, slowed down by a capacitor in the metering circuit, but further slowing would result in an unacceptable lag between an adjustment on the radio and the resulting meter indication.

When the SINAD meter is used as a receiver alignment aid, the amount of flicker can be greatly reduced by the use of an auxiliary filter circuit connected between the loudspeaker terminals and the SINAD meter input. This filter circuit reduces the lower frequency noise components, which contribute most of the flicker. When the circuit is in use, however, the SINAD meter calibration should be considered only relative and the filter should be removed for any quantitative measurements.

6 *SINAD for AM receivers*

The foregoing sections on sensitivity (1–5) apply also to AM receivers, except that signal/noise is measured at 10 dB using a modulation depth of 30 per cent.

Quieting

This applies to FM receivers which use a limiting IF, which is the most common type of receiver in current use. Quieting is largely a function of the IF stages and their ability to limit cleanly and quickly when receiving a coherent signal. The usual quieting figure specified is '20 dB quieting' and is measured as follows:

1 Set up the receiver with no signal entering the antenna connector, with any mute circuits disabled and with the volume control advanced such that some measurable noise is present across the audio load, but that the audio amplifier is not in itself limiting. Measure the noise output.
   Note that the speaker may be replaced by a resistor of equal impedance and sufficient power rating. The audio power can be measured by means of an audio milli-voltmeter calibrated in dB.
2 Introduce unmodulated RF carrier from a test signal generator at a very low level (e.g. about −125 dBm) to the antenna socket. Gradually increase the signal generator level. The noise output from the receiver should begin to fall as the receiver quietens. Keep monitoring the noise output without adjusting the volume control and, when the noise level falls by 20 dB, take a note of the carrier power required to achieve this. The limits on this figure will have been set by the designer and will be found by reference to the manufacturer's specification.

It is possible for a receiver to fail this check and yet have passed the 12 dB SINAD check. A failure on the quieting check indicates low mixer or IF gain. Much of the IF gain on modern receivers is provided by a single IC which often also contains a second local oscillator, mixer and second IF amplifier as well as the FM discriminator and mute circuitry. However, the first mixer, IF filter and amplifier are usually outboard from the IC and are likely to be the most likely cause of the problem.

Ultimate signal to noise ratio

This check is to verify the performance of the modulation detector and audio stages. It has relevance to AM and to FM receivers. It is measured as follows:

1 Apply an RF signal to the antenna socket modulated to full system deviation (for FM) or 50 per cent modulation depth (for AM). Use a carrier level of about −60 dBm which should be substantial enough, i.e. if increased in level further it will have no effect on the noise output from the receiver. The level should not be so high that it causes the front end to limit.
2 Set the volume control such that the receiver is delivering the specified maximum audio output level into the rated load. It is probably worth performing a distortion check at this point. Switch off the modulation and measure the noise level at the receiver audio output. The ultimate signal to noise is the ratio between these two levels and is usually quoted in dB.

There are three points here worthy of note. The first is that ultimate signal to noise is the ratio of noise level to the audio level at full system modulation, whereas for sensitivity checks 60 per cent of full system modulation is used. The second is that full specified audio power is implicated and so this check is embodied in the ultimate signal to noise check. The third point is that often it is the signal generator itself which puts a limit on the ultimate signal to noise. Modern low cost synthesized generators typically return figures of 40 dB or so when used as the signal source for this check, but a good receiver may be 10 dB or more better. One way round having a very expensive low-noise generator is to build a signal source for the check consisting of a crystal oscillator, buffer and attenuator, running on the receiver IF frequency. A suitable design is discussed in Chapter 9. A signal generator must be used to check the achievement of specified audio level, and then the crystal source substituted to measure the noise level, injected into the first IF filter.

This oscillator will need calibration against a frequency counter and an RF millivoltmeter. It is also very suitable as a crystal marker oscillator for zero-beat checks detailed later.

IF bandpass shape

Data is increasingly carried over mobile radio systems and the pick-off point to the data demodulator from the receiver is usually before any AF filtering such as de-emphasis. Correct IF alignment in a well-designed receiver should result in best 20 dB quieting sensitivity combined with a good 12 dB SINAD sensitivity, but a range of IF alignment settings may satisfy this criteria without achieving the best possible distortion figure for, say, a 1 kHz tone at 60 per cent system deviation. If the radio is to be used for data, aligning the IF filter matching circuits while sweeping the IF is highly recommended. A block diagram of a suitable setup is given in Chapter 9.

The sweep rate is limited to about 10 Hz to avoid overshoot of the signal generator, and the amplitude of applied RF signal should be such that a 20–30 dB dynamic range is available on the CRO without the first IF amplifier going into limiting. The amplitude is best found by trial, and when correct, a 2 dB ripple across the passband should be discernible. A slower sweep rate may not be suitable as the external FM input of most signal generators is not DC coupled to their phase lock loop synthesizers. It is worth checking the lower modulation frequency limit in this respect on the generator in use.

Transmitter tests

Measuring output power

The power output of a transmitter is perhaps the most 'looked at' figure by

Mobile Radio Servicing Handbook

misinformed customers; in fact a 0.5 dB drop in receiver sensitivity is deemed insignificant, but the same drop for transmitter power is about 3 W (from 25 W), a decrease considered unacceptable. Transmitter power is affected very much by DC power supply voltage. Ideally the bench regulated power supply sense wires should be connected right at the transmitter power input connector to compensate for volt drop along the power wires in between. If this is not possible then as heavy cable as is practicable (32/0.2 mm for 5 A for example) should be used to minimize the volt drop, and then the power supply voltage set such that the nominal operating voltage appears at the power connector to the transmitter on load. Power is of course proportional to V^2, which accounts for the sensitivity of transmitter power output to supply voltage.

It is also necessary when aligning some PA stages that 150–200 per cent of rated transmitter current will be taken to effect correct alignment. This is because when aligning stages prior to the output stage, the alignment criteria is often maximum current. The transmitter efficiency is largely a function of the PA matching to the antenna and so, depending on exact alignment procedure, a large current may be taken during alignment.

The power specified is usually a minimum figure, and as long as this is achievable using the suggested alignment procedure, the transmitter RF stages are probably all right.

Current taken

A transmitter will generally have a maximum current specification for a given RF power output. This ought to be less than its typical maximum power. For example the specified power may be 25 W when the maximum available at nominal supply voltage is typically 32 W. It is usual to have a margin of this amount so that the power control loop has some dynamic range.

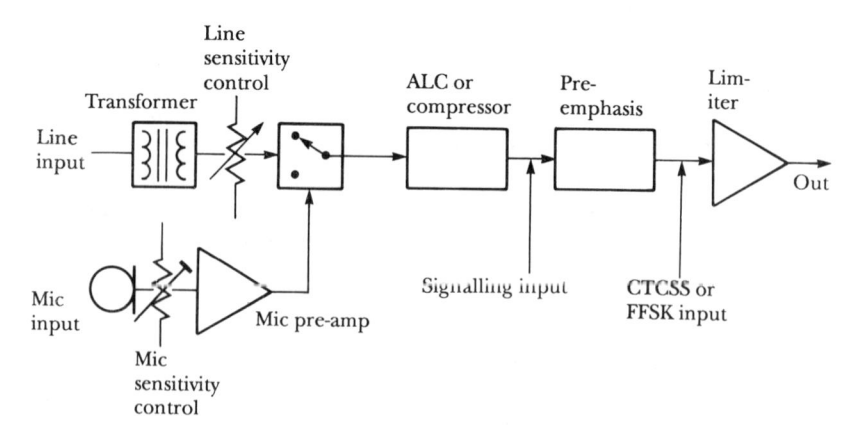

Figure 8.16

Setting AM or FM modulation depth

The following procedures are valid for both FM and AM transmitters, differences where they occur are stated in the text.

The method of setting up modulation depends upon what audio processing system is used prior to the modulator. The block diagram given (Figure 8.16) illustrates a typical configuration.

The transmitter may have a microphone input, a 600 ohm line input, a single-ended data or signalling tone input, or perhaps all three. The procedure suggested here (based on type approval methods) is a good example of modulation setting and is outlined as follows:

1 Connect an AF signal generator across the microphone or 600 ohm line input terminals. The generator output impedance should be equal to 600 ohm or to whatever transmitter input port impedance is specified. If the transmitter has a sensitivity adjustment on this input, it should be adjusted for maximum sensitivity.

2 Set the generator level to 20 dB above the specified level required at that particular level input to give full system modulation. Sweep the signal frequency over the baseband range (typically 300–3400 Hz) to find the frequency at which maximum modulation is achieved.

   Switch the transmitter channel selector through all channels to find the channel of maximum modulation depth. If a channel of 'highest modulation' is found the following procedures should be performed on this channel. If all channels are of equal modulation depth the tests may be performed on any channel.

3 Adjust the transmitter 'hard limit' or 'compressor' stage to give full system modulation which may well be 5 kHz or 2.5 kHz peak for FM or 80 per cent for AM.

4 Set the AF signal generator to 1 kHz and to the level specified for modulation sensitivity. Check that the transmitter audio input sensitivity meets the specification. In the case of a line or a microphone input there may be a sensitivity control and this should be adjusted for maximum for this test.

5 To set the sensitivity control, offer the microphone to a suitable sound box, or connect the line to an AF generator at a specified level, and adjust the control to give the modulation required. It is usual to adjust sensitivity controls to achieve 60 per cent system deviation (FM) or 30 per cent modulation (AM) for a given sound pressure or input level at 1 kHz.

Checking for harmonics and spurious

Spectral purity, a spectrum analyser or tunable calibrated receiver is required for this check. Type-approval authorities place limits on transmitter spurious outputs and the allowed limit has been dropping over the years. At present it is $0.25\mu W$ for some specifications and $2.5\mu W$ for others. It needs a fairly expensive instrument to read such levels when the transmitter power is 25 or 50 W, but one solution is to place a very sharp notch filter between the transmitter and analyser to attenuate the on-frequency signal, thus reducing the dynamic range requirement of the measuring instrument. A typical system is illustrated in Figure 8.17.

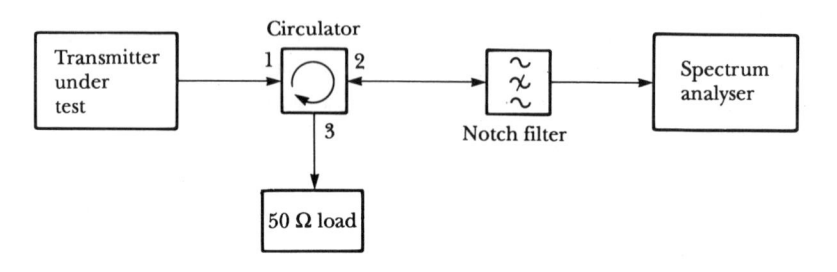

Figure 8.17

The circulator is necessary since the filter operates by mismatching the coaxial system at its resonant notch frequency which results in reflection. The power is then absorbed by the load connected to the third circulator port.

Modulation purity, modulation distortion and signal/noise checks will test correct modulator and signal source operation, whether it is a crystalled or synthesized radio. If the crystal oscillator is noisy or if the phase lock loop synthesizer is unstable, then these will show up as residual, unstable or noise modulation. The modulated carrier will need demodulating first and the AF output from the modulation meter applied to a distortion analyser and/or AF milli-voltmeter. There are a few points to note.

The modulation meter itself must produce noise and distortion an order of magnitude less than the transmitter, so as not to affect result.

The AF level into the transmitter should be equal to the specified sensitivity to give full system modulation depth and the transmitter adjusted to give full system modulation depth.

The check is usually done at 1 kHz but if data is handled, the check may be done at the signalling tone frequencies, such as 1.2 kHz or 1.8 kHz.

The bandwidth in which the distortion or noise products are measured should be the normal baseband of the system, for example 300–3400 Hz. A filter needs inserting between the modulation meter and the AF metering device if these are wideband. The reason is that the modulator produces distortion products as harmonics of the input tone frequency, and the oscillator (or phase lock loop) produces noise, both without definite band limit. It is therefore necessary to define the measurement bandwidth.

CTCSS modulation purity is important, since harmonics of the sub-audible tone will, if above about 300 Hz, propagate through the high pass speech/

CTCSS filter at the receiver and be heard by the user. The phase modulators used in crystalled sets are the main culprits, since as the modulating frequency decreases, the index increases, as does the phase shift required of the modulator. The level of audio signal voltage applied to the modulator increases as the frequency decreases at a rate of 6 dB/octave in order to achieve a true FM response.

For example:

Frequency deviation required	500 Hz pk over the range 60–250 Hz
Modulation index range	8.33–2.0 (index = deviation/frequency)
Phase modulation required	8.33–2.0 Rads

CTCSS modulation places a greater demand on the transmitter than baseband modulation; even though the lowest frequency in the baseband is, say 300 Hz, with pre-emphasis the transmitter is never going to be called upon to produce full system deviation at this frequency.

A single modulator stage can give a maximum phase shift of about eighty-five degrees, which is roughly 1.5 Rads. We need over 8 Rads and the problem is solved by a combination of two or modulator stages in series and by modulation at the crystal generation frequency, then multiplying this up to the final RF frequency.

For example:

Phase modulation required	8.33 Rads.
	Final RF frequency 85 MHz
Crystal frequency multiplication	× 3
Phase modulation required of modulator	8.33/3 = 2.77 Rads peak
Amount supplied by each modulator stage:	1.5 Rads peak to peak
Number of modulator stages required:	4

Note that the number of stages is 4 and not 2, since 8.33 Rads is required in both directions about the carrier.

We assume therefore that in the majority of FM crystalled sets there will be more than one modulator stage and these will possibly operate in pairs, one to deviate the carrier positively, the other to deviate it negatively. If the pairs of modulators do not track accurately, this will introduce distortion, and the larger the input audio signal, the worse the distortion becomes. So the CTCSS tone will suffer the greatest.

With synthesized FM sets, the phase lock loop will have been designed to cope with CTCSS modulation, i.e. its bandwidth chosen so as not to wipe it off. Speed up circuits are generally used to overcome the resulting inherent slow lock time. It is to the speed up circuits and the loop filter portions of the synthesizers that suspicion should fall if CTCSS modulation is a problem.

It is unusual to fit or find an AM transmitter with CTCSS. Usually a transformer is used to modulate the RF stages and the core non-linearities are too great if the transformer is to be a sensible size.

9 Calibration, test and fault finding

Test equipment requirements and use

A range of test equipment required to set up and maintain mobile radio equipment is listed below.

1. An RF signal generator capable of AM and FM. CTCSS encode plus a 1 kHz tone is useful.*
2. A SINAD meter, either to read 12 dB SINAD (flat) or 20 dB SINAD (CCITT weighted).*
3. A signal/noise or AF power meter to read −30 dBm to +20 dBm across 600 ohm. A switched internal load is useful to allow monitoring a line without loading it.*
4. A 10.7 MHz/21.4 MHz crystal marker oscillator.
5. An RF power meter to cover 5 W FSD to 100 W FSD over the frequency range 50 to 500 MHz.*
6. A modulation meter capable of AM and FM. AM ranges of 50 per cent and 100 per cent FSD and FM ranges of 1 kHz, 3 kHz and 5 kHz deviation. A switched low-pass filter at 300 Hz is useful for measuring CTCSS deviation.*
7. A 10 A and 20 A FSD DC current meter.
8. An AF signal generator to deliver −30 dBm to +10 dBm into 600 ohm to cover the frequency range of 50 Hz to 10 kHz.*
9. An AF/RF frequency counter 50 Hz to 500 MHz.*
10. A dual-trace CRO capable of X<>Y switching for Lissajous patterns.
11. A bench power supply capable of delivering 13.8 V at 8A maximum for 25W transmitter mobiles, and 15 A maximum for 50 W transmitters. A ±20 per cent variability on the power supply could aid fault finding, and 24 V and 50 V sources are useful for checking 24–12 V convertors and telephone circuits related to fixe stations.

All items marked * are available together in dedicated mobile radio test sets such as the RTT Comtest, Marconi TF 2955 Solartron 4040, Rohde and Schwarz CMT, or IFR.

In addition to that listed, some self-made equipment will make life easier. Two examples are a diode-probe detector for peaking resonant circuits, and a

Mobile Radio Servicing Handbook

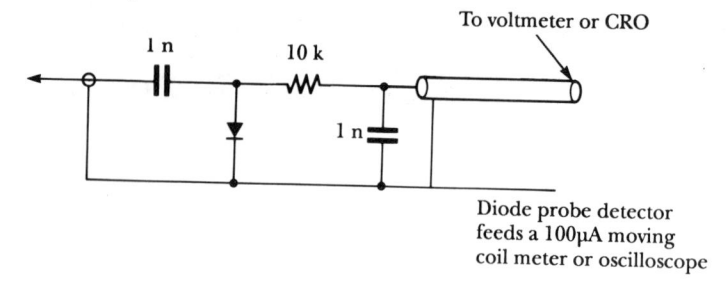

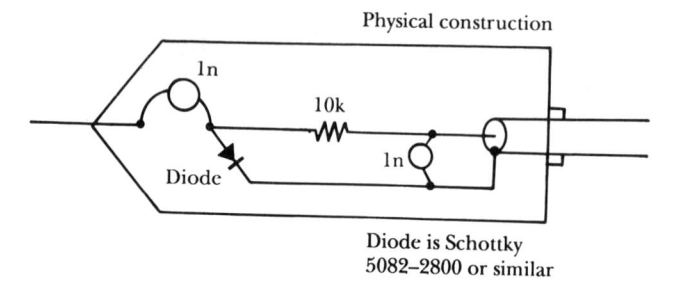

Figure 9.1 *Diobe probe detector – this can feed a 100 µA moving coil meter*

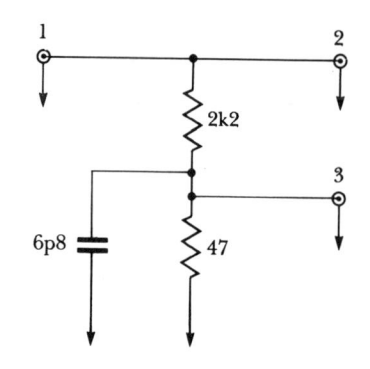

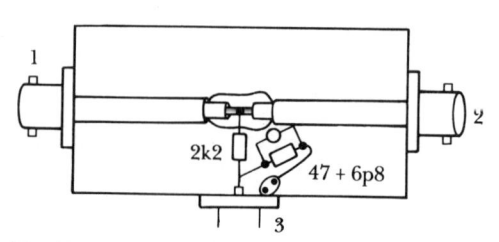

Figure 9.2 *40 dB sniffer – construct in metal box, break jacket and screen in 'through' path with as little gap as possible*

Calibration, test and fault finding

'sniffer' to enable monitoring of a transmitted RF signal roughly 40 dB down, so as not to damage a modulation meter or spectrum analyser. Suggestions of circuits and mechanical arrangements for these are given in Figure 9.1 and Figure 9.2.

The diode probe detector uses a hot-carrier diode to rectify the RF signal and fits neatly into a felt-tip pen case. It is necessary to ground the probe only to prevent hum pick-up when used with a CRO.

The sniffer is a simple resistive divider network and the purpose of the shunt capacitor across the output is to keep the potential divider ratio roughly constant over the UHF region where the signal begins to 'hop' over R1. Both circuits can be effectively used up to 500 MHz.

A useful piece of test equipment to perform the ultimate signal-to-noise ratio test is a low noise oscillator. A suitable circuit is shown in Figure 9.3, and the oscillator, once tested, can be built into a diecast box to minimize external radiation. The oscillator runs at the receiver IF, and should be injected directly into the first IF amplifier. An ultimate signal/noise test checks the demodulator and audio stages of a receiver.

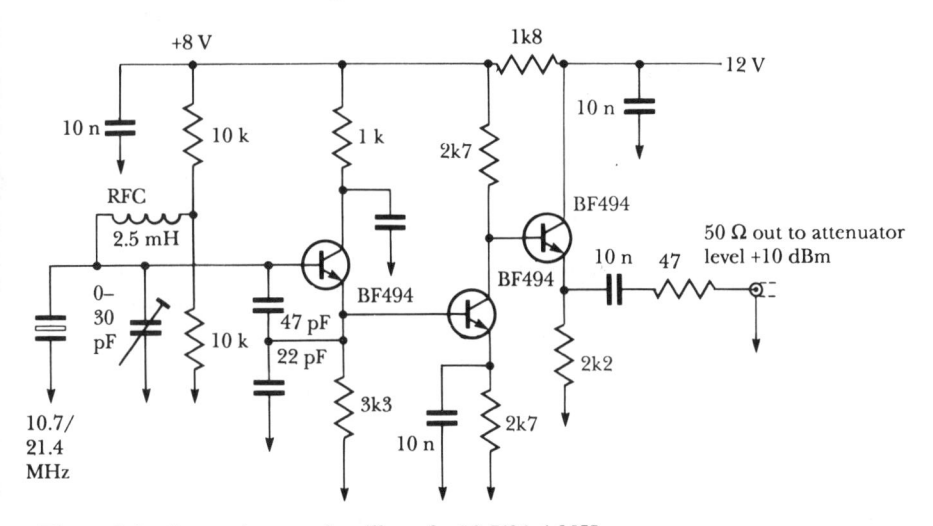

Figure 9.3 *Low noise crystal oscillator for 10.7/21.4 MHz*

The pitfalls encountered in using test equipment are mostly due to finger trouble, some common ones are considered below.

1 If the SINAD or AF millivoltmeter is included in another piece of test equipment, make sure the loudspeaker terminals are connected to it in the correct polarity. This rather obvious point often catches people out and the usual symptom is AF instability due to the ground return loop.
2 If the external modulation input to an RF generator is used, make sure the modulation is what it says it is on the front panel meter. Some generators are

designed to be accurate with only one input level. Check the generator output on the modulation meter.

3 Don't connect the output of a transmitter directly to a modulation meter, counter, spectrum analyser etc. A rather obvious point, but everyone 'squirts' a piece of test equipment at least once: one cannot be too careful. Use a 'sniffer' or power attenuator to reduce the level to $<+10$ dBm.

4 Some modulation meters and RF signal generators are very susceptible to external fields from mains transformers resulting in residual 50 Hz modulation indication. Moving the instrument a few centimetres will identify this.

5 Some RF generators have direct case radiation which upsets receiver sensitivity measurements. The clue here is whether inserting more attenuation in the coaxial path between the generator and receiver results in the expected worsening of audio quality from the receiver. If not, then the chances are that the receiver is picking the signal up across the bench directly from the generator case; the only cure is to improve the screening of the generator or increase the distance between the two.

6 Make certain that good quality test leads are available. 0.5 dB loss in a transmitter RF lead means a drop of about 2.7 W from a 25 W transmitter. A good quality 1m lead should have about 0.2 dB loss.

Alignment and calibration

Before delivery to the end user

It is possible that radio equipment will go straight from the production environment to the end user, but it is more usual for another department or company (such as a dealer) to handle the goods and customize them by, for example, putting them on the required frequency or fitting signalling options. Time is normally at a premium at dealers' premises so the tests are kept to a minimum.

1 *General*

It may be necessary to run the set up from scratch if suspected that either it is not new or that a test at the factory was not performed. In either of these instances it is better to follow the procedures laid down in Chapter 8.

If a factory test has been performed, there may well be a test report with it giving the frequency of run-up and the results achieved, and it can be assumed that upon switch-on nothing will explode. The problem then reduces to retuning the set on the desired frequency. Note that in all cases, the golden rule is to follow the manufacturer's instructions if these are available. If not, then the following may be helpful when aligning tuned circuits.

If an inductor core has two positions within the former where a peak is found, then the tuned circuit is probably resonating at the desired frequency on either peak. If, however, there is a single peak and the core is centrally

Calibration, test and fault finding

positioned in the coil or is hanging out of one end of the former, then more or less turns respectively need connecting into the circuit; alternatively more or less capacitance is needed which is associated with that inductor.

If adjustments consist of variable capacitors it is likely that one fixed small capacitor will have to be added at the bottom end of the frequency range. Look out for this: such a capacitor may have to be added or removed if going to or from the bottom of the band. Variable capacitors typically have a max:min ratio of 5:1 whereas slug-tuned RF inductors have about 2:1. Therefore, a variable capacitor can swing a resonant circuit further than a variable inductor. Obviously there will only be one tuning peak associated with a capacitor, and visual inspection should reveal whether the plates are a desirable 30 to 70 per cent meshed. If a moulded plastic case variable ceramic capacitor is encountered, it should exhibit two peaks per complete revolution.

2 *Tuning the receiver*

It cannot be overstressed that the best procedure to follow is that laid down in the handbook by the manufacturer. However, this basic procedure should be successful.

a) Setting up the local oscillator

Crystalled sets: Calculate the correct crystal frequency. There will be a formula for this, but if unknown and impossible to obtain, it is possible to substitute the local oscillator with a second RF generator connected to the oscillator input port of the mixer. The local oscillator frequency will be spaced from the RF input frequency by the IF, but the direction of the spacing may be unknown. This can be determined later, but in order to determine this the front-end RF tuned circuits must be aligned. So set the second generator to either high or low side, adjust the RF filters between the antenna input and the RF amplifier and between the RF amplifier and the mixer, and we shall return to the local oscillator afterwards.

Where the crystal generation formula is known, the local oscillator chain is adjusted for maximum oscillator level at the correct frequency into the mixer. Note that this may not mean maximum sensitivity; sometimes the local oscillator chain is finally adjusted to give maximum mixer current. This is particularly true of base station receivers with a 'strong' mixer. If this is the case, then a test point will be provided to connect a DC meter. Test point locations provide the clue to local oscillator alignment. Often they will be found on transistor bases following resonant circuits to enable a diode- probe or RF voltmeter to be attached and the resonant circuit tuned.

To put the local oscillator on frequency, it is possible to adjust it against a crystal marker oscillator running on the IF frequency. The crystal frequency trim component is usually across or in series with the crystal and may either be an inductor or capacitor, usually adjustable. A beat note is heard in the loudspeaker and the crystal is trimmed for zero-beat. This is accurate only to about ±100 Hz since the responses of the audio stages and the ear fall off below this, but this is usually adequate. If this is difficult, a spectrum analyser can be used to pick up some local oscillator signal from the antenna terminals, and the unmodulated output from an RF generator mixed with it for comparison on the analyser screen.

If the zero-beat method is used, obviously some IF signal must be present, so the front-end filters must be somewhat on tune. Typically, with correctly aligned front-end filters, about −80 dBm of signal input (unmodulated) will give a perfectly audible beat note. If the front end is off tune then more will be required. Fig 9.3 is suitable also as an IF marker oscillator.

Synthesized sets: Programme the synthesizer. All modern frequency synthesis uses dual modulus prescaling in association with counters programmed by EPROM, diode matrix or switches. The counters are usually on an LSI CMOS device and are programmed either in parallel or serial format from some form of lookup table whose address is determined by a channel switch on the front panel. In manuals for synthesized radios there will be charts for programming the EPROM, diode patterns etc., and it is no easy task to program these radios without the proper manual.

If the instructions are not to hand, then, in summary, the following needs to be known:

- VCO output frequency required;
- reference frequency, i.e. frequency at which phase detection is performed;
- prescale modulus numbers.

The reference frequency is typically equal to or a fraction of the channel spacing of the radio (e.g. 12.5 or 6.25 kHz) and is related to the VCO output frequency by the programmable counters which together with the prescaler, form the total divide ratio.

This total divide ratio is

$$N_t = VCO/Ref$$

and is dimensionless and of the order of tens of thousands. N_t is related to the prescaler modulus numbers and the programmable counter numbers by

$$N_t = NP + A$$

Where N_t = total divide ratio, N = N counter (the larger), A = A counter, P = prescaler modulus (lower number).

From these formulae it is possible to arrive at the numbers which must be programmed into the two counters. The method from here on is product specific and may, for example, entail converting binary programming patterns to hex code for EPROMs or calculating out the absolute meaning of each programming wire in terms of VCO frequency shift to arrive at a switch or diode pattern.

Calibration, test and fault finding

There will normally be a method of introducing loop error into the synthesizer in order to set a voltage at a test point, usually the VCO control voltage. A variable reactive component in the VCO will vary the VCO control voltage by the loop action and it is usual to adjust this to achieve a VCO control voltage midway between the supply rails of the amplifier providing the voltage. This gives the synthesizer the maximum dynamic correction capability. If the radio is multi-channel then a variety of VCO voltages are encountered and these should be placed symmetrically about the midway voltage.

b) Adjusting the front-end RF tuned circuits

The test equipment configuration is shown in Figure 9.4

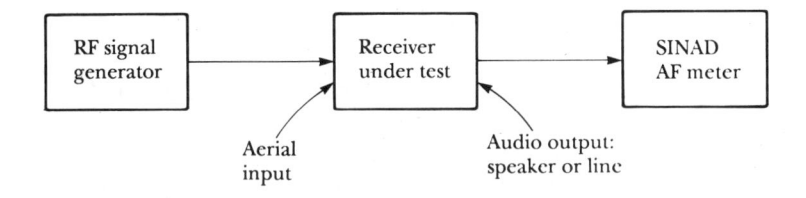

Figure 9.4 *Test equipment set up for tuning receiver front end*

It is usual to tune to maximum sensitivity. Stagger tuning is rare in mobile radio, however there may be notch filter 'traps' for the image response frequency (RF frequency ±2 × IF / high or low side injection respectively), or even for the IF to prevent radiation of this from the antenna. Two typical

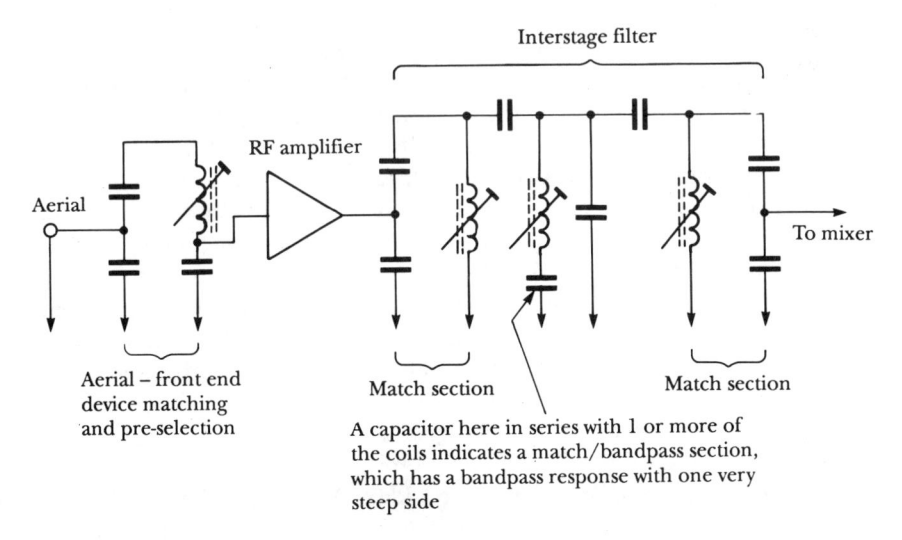

Figure 9.5 *Typical VHF front end filter configuration*

117

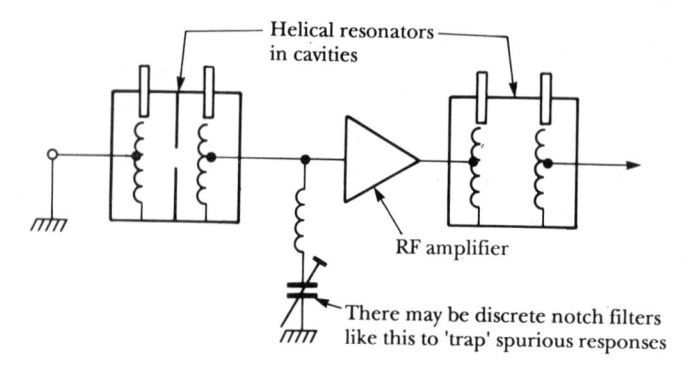

Figure 9.6 *Typical UHF front end filter configuration*

filter configurations are given above; one for VHF receivers (Figure 9.5) and one for UHF receivers (Figure 9.6).

For FM receivers, the parameter to test for is SINAD sensitivity. Traditionally, receiver sensitivity is quoted as a certain minimum RF level into the antenna terminals carrying 60 per cent system deviation, to achieve 12 dB SINAD at the audio output terminals. This is measured in a flat 300–3400 Hz bandwidth. Recent thinking on data link type receivers has resulted in a 20 dB SINAD sensitivity being quoted which is measured using a psophometrically weighted filter in the audio path, according to CCITT recommendation P.53a. This has relevance where there is little audio processing (e.g. no de-emphasis). Subjectively the two are the same; 12 dB SINAD (unweighted) is approximately equal to 20 dB SINAD (weighted).

While adjusting the filters for maximum SINAD sensitivity, keep the generator level at about the 12 dB SINAD sensitivity level, i.e. back off the level as each tuned circuit is peaked. On most SINAD meters, the scale is most useful between 6 and 12 dB SINAD, and so it makes sense to keep adjusting the generator output level to stay within this range. It is worth checking at least once during this procedure that the local oscillator is on frequency; if wildly out ($>\pm 2$ kHz, say) then a poor SINAD reading will be obtained at all times due to excessive distortion. See the section on local oscillator adjustment above. A typical result should be about -119 dBm for a normal receiver with de-emphasis, or about -113 dBm for one without de-emphasis. These figures are approximate.

For AM receivers, the same basic procedure applies except that the sensitivity parameter is 10 dB signal/noise; AM receivers are not as sensitive as their FM counterparts, a typical sensitivity being -107 dBm with thirty per cent modulation at 1 kHz. A simple AF millivoltmeter may be used to measure signal + noise/noise. The difference between S + N/N and S/N becomes insignificant for our purposes when the ratio is 10

Calibration, test and fault finding

dB or higher. AM receivers do not have limiting IF stages and so the front-end noise is not present at the audio output in absence of signal. As input signal is applied gradually from zero, the noise and signal rise together but the noise rises more slowly up to a certain point, then the noise reduces, hence the S/N is improving. This characteristic is dependent upon the AGC performance and maximum available RF and IF gain, and varies amongst receivers.

c) If the crystal generation formula is not known, then the second generator is still being used as the local oscillator. Try altering the generator frequency to compare 12 dB SINAD/10 dB S/N sensitivities both when it is high side and low side. Go for the best one. If both give the same result, then the safest thing to assume is that the oscillator runs high-side for frequencies up to 100 MHz and low-side above this. The reasons for this are connected with typical front-end filter performance in suppressing spurious responses. In order to determine the multiplier number, connect the generator across the crystal terminals and try numbers from 2–6 (VHF sets) or 6–36 in steps of 3 (UHF sets), and see which gives the most sensible settings of the resonant circuits in the local oscillator chain. When connecting the generator, it is safest to use a blocking capacitor so as not to upset the circuit DC conditions. This procedure can be lengthy and is a poor substitute indeed for knowing the correct formula.

d) IF circuits

There is not normally any adjustment to be made, as the IF section is lined up at the factory. However, if an IF filter is changed during, for instance, a channel bandwidth change then some peaking is necessary. The safest procedure is to adjust the IF while sweeping its response, as a flat response (to ± 1 dB) over the IF passband indicates correct alignment. However, adjusting the IF section for maximum SINAD sensitivity will give acceptable accuracy for speech. For data traffic and AM sets though, it is recommended that the IF be swept.

e) AF circuits

The only adjustments necessary here are to set the audio levels to the customer requirements. If line levels are involved, the line launch level specified by BT is -13 dBm. It is worth viewing the output waveform at say 1 kHz on an oscilloscope with a fully quieting signal and 60 per cent system deviation (FM) or 30 per cent modulation (AM) as a quick check that it is a reasonable sinusoid. For AM receivers, a check should be made of the AGC operation. For a given rise in modulated RF input signal, the audio output level should rise by a small limited amount. The rise in audio output is typically 1 dB for a rise in RF input from about -107 dBm to -27 dBm.

Transmitter tuning

a) Setting up the oscillator

Crystalled sets: Calculate and fit the correct frequency crystal. If the crystal formula is unknown, then it will be necessary to substitute the crystal for a signal generator and to monitor the oscillator stage output on an RF millivoltmeter. Take an inspired guess at the multiplier number; for VHF transmitters it is likely to be 2, 3, 6, or 12, and for UHF transmitters 12, 24 or even 36. Figure 9.7 gives the general idea.

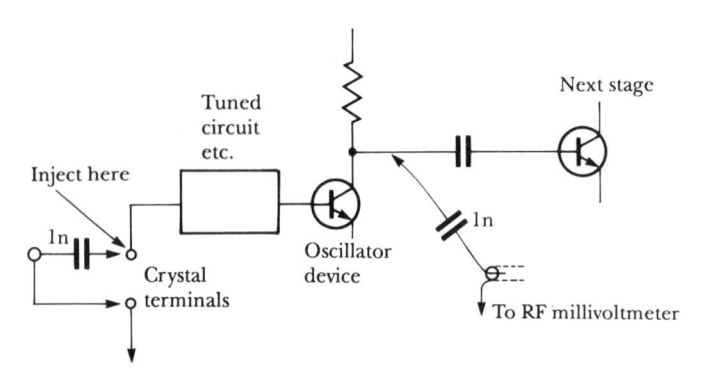

Figure 9.7 *Checking optimum frequency for oscillator circuit*

The circuit will probably have greatest amplification at the correct frequency. An alternative method with a series mode oscillator is to connect a resistor of around 27R with a 1n blocking capacitor across the crystal terminals. This will look like a series resonant crystal to the circuit and hence it will oscillate at a frequency determined by the circuit tank. Thus by counting this frequency and adjusting the tank circuit, the frequency range of the crystal is measured. If neither method yields results, then stick to the generator substitution method and attempt to tune the multiplier chain, being prepared to try various multiplier numbers to optimize levels further along.

Synthesized sets: The same general comments apply here as to receiver synthesizers in programming and setting up. The VCO will normally run on the final frequency, that is, no multipliers are used in synthesized radios. The modulation is applied directly to the synthesis loop and is true FM as opposed to phase modulation normally encountered with crystalled transmitters. Separate VCOs may be used for the transmitter and receiver. This has advantages such as faster loop lock time and the absence of modulating components in the receiver local oscillator.

b) Aligning the stages to the exciter output

This section is equally applicable to FM and AM transmitters. Normally the transmitter is divided into two identifiable parts – exciter and PA. If the two are connected by a coaxial cable, then the impedance between the two will be 50 ohms and direct connection of a power meter to the exciter output is possible. In crystalled sets there may be several resonant circuits to tune, and this is best done with the help of a diode-probe or RF voltmeter

Calibration, test and fault finding

monitoring the output of the stage after that which is being adjusted. Try to have one stage in between the probe and adjustment; the stage will buffer the probe capacitance from the resonant circuit being tuned. The modulator stage(s) will directly follow the oscillator stage followed by multiplier and amplifier stages. A typical UHF arrangement is illustrated in Figure 9.8.

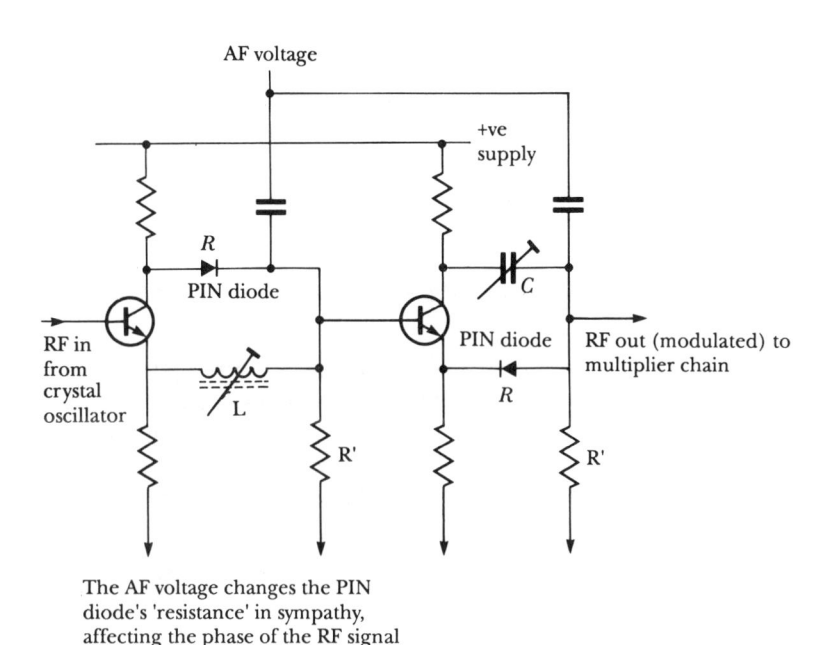

The AF voltage changes the PIN diode's 'resistance' in sympathy, affecting the phase of the RF signal

Figure 9.8 *Typical phase modulator circuit for FM (UHF)*

Again stagger tuning is rare and the procedure is to tune for maximum level at the correct frequency. Watch out for resonant circuits with many tuning peaks; it is possible that the swing available in some will encompass incorrect harmonics and is shown up by nearly impossible tuning of the next stage. In cases of extreme difficulty, a spectrum analyser may be connected to determine whether the resonant circuit is indeed tuned to the correct crystal harmonic.

Synthesized sets have little, if any, tuning as far as the exciter output, as wideband operation is possible, i.e. no harmonic selection is required. This also means that most stages are amplifiers working in Class A, with perhaps only the exciter output stage working in Class C. By contrast, crystalled sets have non-linear amplifiers to generate harmonics for later selection by filters. If there is any tuning required, it is to peak the exciter output on the frequency used. A typical UHF synthesized arrangement is illustrated in Figure 9.9.

Mobile Radio Servicing Handbook

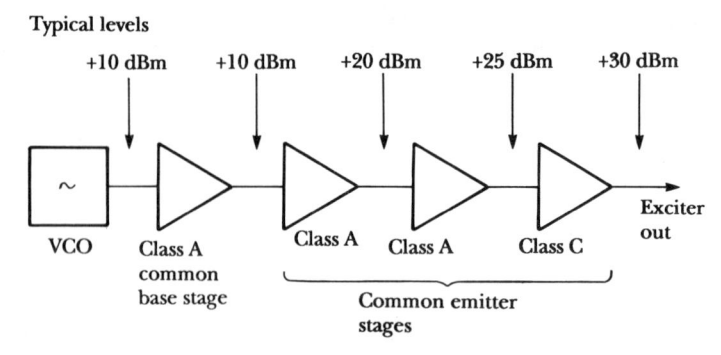

Typical levels

+10 dBm, +10 dBm, +20 dBm, +25 dBm, +30 dBm

VCO → Class A common base stage → Class A, Class A, Class C common emitter stages → Exciter out

A common base stage is usually included as its reverse transfer coefficient is zero. If not present, VSWR changes at the antenna would upset the VCO!

Figure 9.9 *A common UHF RF amplifier configuration for a synthesized transmitter exciter*

Between 0.3 and 5 watts is a typical exciter output power, and as said earlier, may be measured directly at a coaxial link between the exciter and PA. Generally, VHF PAs are higher gain and thus require less drive. If there is no such coaxial link, then proceed directly to setting up the PA. However, some idea of exciter output stage performance can be gleaned from the current it draws. If the PA is disabled by open-circuiting the base or gate connection of the PA driver, the principal consumer of current is

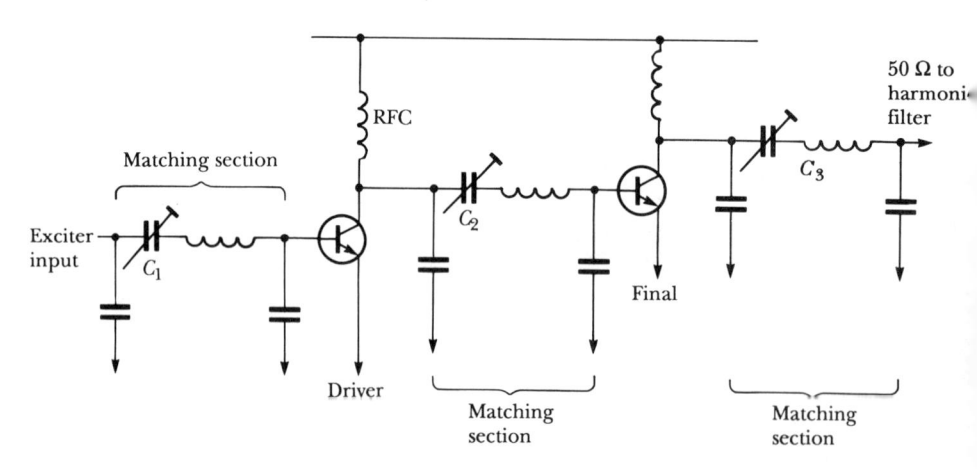

Set C_1 and C_2 for maximum PA current and adjust capacitors after the final stage (e.g. C_3) for maximum efficiency at the rated PA output

Figure 9.10 *A typical PA configuration*

Calibration, test and fault finding

the exciter output stage, and depending upon power delivery may range from roughly 150–500 mA.

c) Setting up the PA

A rule of thumb here is to adjust all variable reactive components before the final stage for maximum efficiency at the PA rated maximum output. If there is a power level control, this is then adjusted to reduce the power if necessary (such as for a CBS trigger station) without further adjusting the PA. This ensures that the PA is working in its most stable matching environment. Figure 9.10 illustrates the point.

PAs up to about 25 W have one device in the final stage, above this two are used, sometimes working in quadrature so that their second harmonics cancel. If two devices are used there are more tuning adjustments but the basic rule of thumb above still applies. Some of the adjustments may interact making readjustment necessary.

Typical currents for PAs are roughly 3 A for 10 W, 5–6 A for 25 W and 9–12 A for 50 W depending on whether it is VHF or UHF. UHF PAs tend to be less efficient than VHF ones.

d) AF stages and modulator

FM: There may be a microphone input, line input, data input, CTCSS tone input or a combination of all four. Base station transmitters will normally have a line input and associated line sensitivity control on the front panel. All four inputs require slightly different audio processing and the best way to look at this is to refer back to the block diagram of a typical audio processor, Figure 6.2.

The microphone is pre-amplified using a low-noise stage and a switch follows this to choose between the microphone and line inputs. Some form of ALC or compressor stage follows this and then pre-emphasis, limiting and an integrator stage. The integrator stage is only used with phase modulated crystalled sets and is used to provide a 6 dB/octave response required by this. The data enters the system just prior to the limiter stage, because, if subjected to ALC or pre-emphasis, corruption will occur. The CTCSS tone input is normally applied to the integrator stage if fitted or directly to the modulator.

A mobile transmitter will not have a line input (unless modified for remote operation) and probably will have a maximum of two adjustments: microphone sensitivity and 'hard limit' modulation depth. A base transmitter may, in addition, have an ALC output control. The correct procedure for setting the hard limit control is:

i Connect an AF signal generator across the microphone or line input terminals, set to 1 kHz at a level 20 dB above the rated sensitivity for that input. A typical microphone sensitivity is 6 mV rms.

ii Key the transmitter and adjust the appropriate sensitivity control to maximum, with the hard limit adjustment at about 50 per cent travel.

Mobile Radio Servicing Handbook

iii Sweep the AF generator frequency over the range 300–3400 Hz and locate the frequency at which maximum deviation occurs. Set the hard limit adjustment to maximum system deviation (usually 2.5 kHz peak for 12.5 kHz channel width, or 5 kHz peak for 25 kHz channel spacing).

Most technicians, however, just whistle or bellow 'four' down the microphone in order to set the deviation hard limit. Maximum deviation commonly occurs with a modulating frequency of 2–2.5 kHz, and few people can whistle up there. As a rule of thumb, if a whistle or a 'four' is used, the hard limit should be adjusted to about 85 per cent of full system deviation, e.g. 2.2 or 4.5 kHz peak. The proper way, and the method used by the type approval authorities, is basically that given in the preceding paragraph.

If an ALC adjustment is encountered, it is set for a given level out of the ALC stage over a wide range of input signals at 1 kHz. A typical output level is 100 m V rms, but check in the manual.

A CTCSS input level adjustment may be provided, and if so it should be set to give the required deviation for the particular input level used. Typical CTCSS deviations are 350 Hz peak for 12.5 kHz channel systems, and 600 Hz peak for 25 kHz channel systems. These vary between manufacturers and opinions are divided as to the optimum deviation.

AM: The audio output stage is normally used in conjunction with a transformer as the modulator stage on transmit. There may be a microphone input, line input and data input, or any combination, each requiring different processing. There is usually a microphone pre-amplifier stage, a switch to choose between the line and a microphone input, an ALC or compressor stage and an output stage. There is no need for a hard limit stage as in the FM case, as increasing modulation depth does not entail more occupied bandwidth. A typical AF processor is illustrated in Figure 9.11.

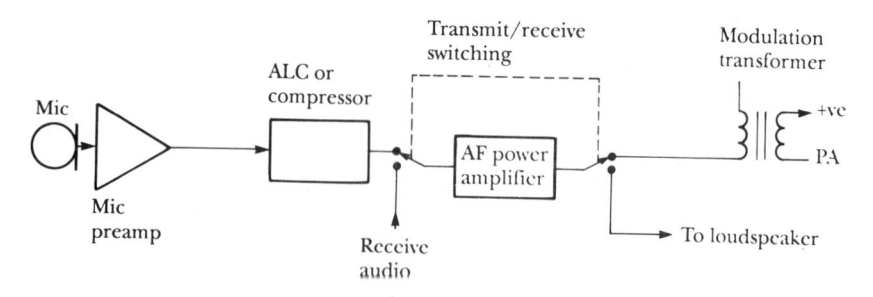

Figure 9.11 *A common AM transmitter audio processor arrangement*

A procedure for setting up the modulation depth is as follows:

i) Connect an AF signal generator across the microphone or line input

Calibration, test and fault finding

on 1 kHz at a level 20 dB above the rated sensitivity for that input (typically 5 m Vrms for a microphone input).
ii) Adjust the ALC/compressor level to achieve 80 to 90 per cent modulation depth.

A whistle or a 'four' bellowed into the microphone will then give typically 50 to 70 per cent modulation depth.

Modern AM transmitters may have integrated AF power amplifiers, but older ones have discrete transistors running in Class B with associated adjustments for quiescent current. If working in the dark, 20 mA per device is a ball-park figure.

In-service fault finding receivers

Introduction

This section deals with fault finding on receivers which have once worked properly, and are thus assumed to have been correctly built in terms of correct components fitted. Faults like trapped wires and dry joints may manifest themselves after a period of service, however the section concentrates on locating the fault to circuit module level and then to component level; this should be valuable in locating dry joints to a rough area but these will not specifically be mentioned. A dry joint will normally give itself away by becoming intermittent when the circuit board is stressed mechanically; it's a case of looking for it.

Receivers

Divide the receiver into blocks:

1 *RF block*
Input filter, RF amplifier, interstage filter, first mixer.
2 *IF block*
First IF filter, first IF amplifier, second mixer, second IF filter, second IF amplifier, demodulator.
3 *LO block*
Crystal oscillator, multiplier chain, synthesizer.
4 *AF block*
Audio stages, squelch.
5 *Power supplies, switching logic.*

The general idea is to locate the faulty block by starting at the audio output terminals and moving back towards the antenna. One can quickly divide the receiver into two major blocks by injecting a modulated signal at the IF into the first IF filter. If the signal gets through from here, but not from the antenna, it rules out the IF and AF blocks and suspicion falls on the RF and LO blocks.

Indeed, the mixer is a meeting place between three of the four blocks listed, and is the most common starting point.

In order to assist locating a faulty block, listed below are typical injection levels required of a modulated signal to the various block inputs for 12 dB SINAD (FM) or 10 dB S/N (AM). AM figures are in brackets:

Injection point	Level
De-modulator IC (FM only)	-80 dBm
IF block	-90 dBm (-80 dBm)
Mixer	-107 dBm (-95 dBm)
RF device input leg	-120 dBm (-110 dBm)

Remember that if the receiver gives an output when an IF signal is applied to the first IF filter, but not when an RF signal is applied to the mixer input, then the LO block becomes suspect. The power supply and switching logic block is shown faulty by wrong DC conditions. On many receivers, the RF amplifier and/or mixer may be fed from the full 12 V supply rail, so they can handle large signals, but the remainder of the circuit is fed from a stabilized supply of say 6 or 8 V. The current drawn from this regulated rail will not be greater than about 50 mA, so the regulating series pass device will not be particularly manly, and therefore susceptible to damage from soldering while the set is switched on, dropping screwdrivers inside, or from circuit faults which cause heavy current to be drawn from the regulated rail.

The various circuit blocks can now be examined to component level detail.

1 *RF block*

All receivers working at about 80 MHz or more will have some form of RF amplifier stage before the mixer. This stage consists of a low-noise amplifier between two bandpass filters. The purpose of the filters is to preselect the range of frequencies over which the amplifier will respond. This is needed for two reasons:

a To prevent unwanted large signals from arriving at the amplifier input so it does not overload and go non-linear.
b To prevent signals entering the mixer which would result in a receiver spurious response such as image or half-IF.

The larger the IF frequency the further away are the spurious response frequencies, and the less selective the RF filtering has to be from that point of view.

The purpose of having a low-noise amplifier before the mixer is to overcome the mixer noise. The local oscillator may be quite noisy, especially if it has been multiplied up from a crystal, and the switching action of the mixer makes a lot of flicker noise. It is always assumed that the noise an amplifier makes happens at its input. Therefore, the RF amplifier under discussion lifts the signal along with its own noise to be at least 10 dB higher than the mixer noise. Then at the mixer output, the noise present is over 90 per cent RF amplifier noise, together with the amplified signal.

Calibration, test and fault finding

In a good design, the receiver noise figure is almost equal to the noise figure of the RF amplifier device, plus the losses before it. Since the losses before the RF amplifier device add to the receiver noise figure, it is usual to have a simple 1-section tuned filter ahead of the device, to prevent high-level unwanted signals from overloading the stage, e.g. the transmitter signal. After the device follows a 3- or 4-section filter where the main selectivity is contained to suppress spurious responses. The loss allowable in this filter is dependent upon RF stage gain, and should not be greater than would prevent the RF amplifier noise being 10 dB or more above the mixer noise. A typical arrangement of gains and losses from the antenna connector to the mixer output is given in Figure 9.12.

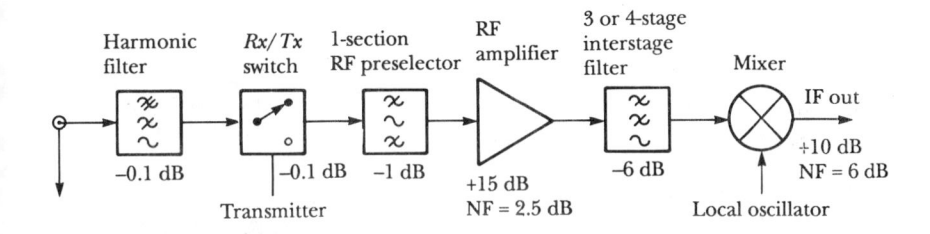

Figure 9.12 *Receiver front end showing losses, gains and noise figures*

The devices used are bipolar or FET in common emitter/common source or common base/common gate. One problem with VHF RF amplifiers is keeping the gain down while maintaining good high-level signal handling ability. One can reasonably predict that devices in UHF and VHF sets from 150 MHz upwards will have common emitter/common source connection, while below this frequency it may be common base/common gate. The four configurations are illustrated in Figure 9.13.

Typical DC conditions are given; supply rails will vary but it is always true that the base-emitter voltage of a bipolar device will be 0.7–0.8 V, and the gate 1-source voltage of a depletion FET will be 1–3 V. The respective collector/drain voltage will be substantially higher than this and supply-rail dependent but typically 6–12 V. All front-end RF amplifiers work in Class A, and are therefore forward biased. Checking of the voltages mentioned will always detect a faulty device. If the voltage is zero then the device is short-circuit, and if too high then the device is open-circuit (this is the more common with static damage such as lightning discharge).

The device could be a 'helicopter' flat package with four wires (two will be the emitter and go right through the device if bipolar), or a three wire ended plastic package, or a four wire metal can package, the fourth wire being a screen if bipolar.

The RF stage in an AM receiver may have its gain reduced by delayed AGC

Mobile Radio Servicing Handbook

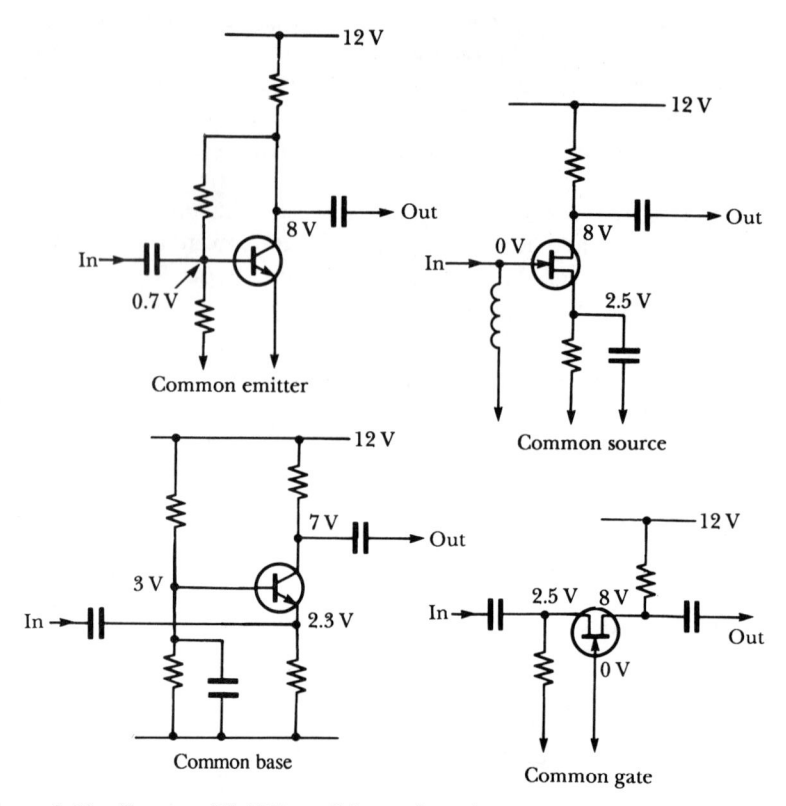

Figure 9.13 *Four possible RF amplifier configurations*

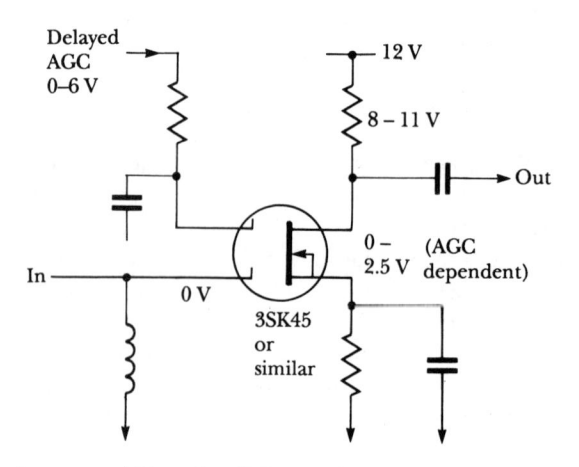

Figure 9.14 *A common AM receiver RF amplifier circuit*

Calibration, test and fault finding

when the signal is very strong and the set is running out of dynamic control of its IF amplifier. AM sets are VHF only and commonly have a dual gate MOSFET in this position. One common configuration is illustrated in Figure 9.14.

The mixer may consist of one or two devices or of a diode ring. A diode ring mixer will have a fixed loss of about 7 dB and require a substantial local oscillator drive of about +7 dBm. However, their intermodulation performance and general ruggedness is better than active mixers. They are also easier to fault find, consisting of the diodes themselves plus two transformers usually wound on ferrite rings.

More usually the mixer is active to overcome the loss and high local oscillator drive problems and may be a balanced pair of devices, particularly below about 225 MHz. Balanced active mixers are difficult to implement at UHF, the diode

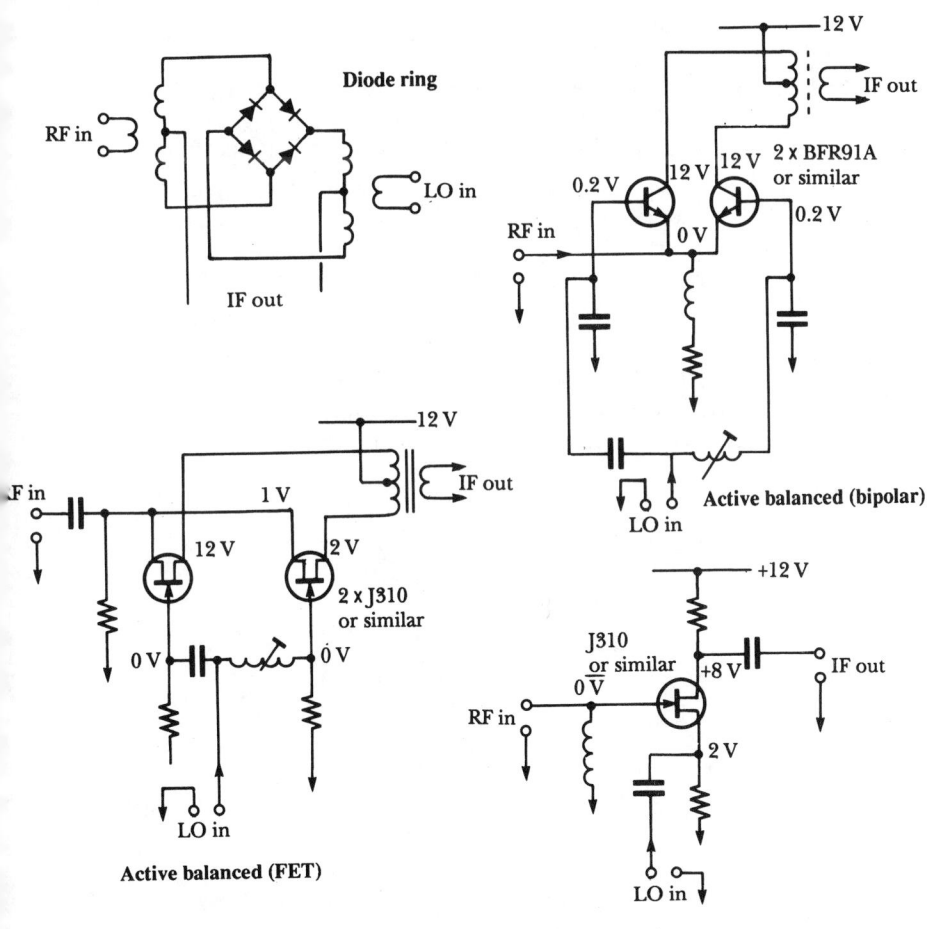

Figure 9.15 *Some mixers*

ring variety is available in UHF packaged in a metal can. To summarize, Figure 9.15 shows some popular configurations with typical DC operating conditions.

Active mixers are usually biased right on the knee of their forward transfer characteristics to make their output current as non-linear as possible when switched by the local oscillator. With a bipolar mixer, the biasing can be arranged with diodes as shown; with JFET the device is biased almost off, and with a dual gate MOSFET the gate 2 voltage is held low.

Measurement of the DC conditions will not necessarily detect a faulty device, since they may work at or near zero bias. A device may well rectify the local oscillator signal and appear to have a negative bias, and yet still be conducting current. The safest procedure is to disable the local oscillator, measure the DC conditions, and if a device is suspect, unsolder all except one leg and check the forward and reverse junction resistances. If a dual gate MOSFET is suspected of static damage, replacement is the only check.

2 IF block

An IF filter follows the mixer in most cases. This is usually working between impedances of several k/ohms shunted by a small capacitance. It will be about a 4-pole filter in a dual IF receiver or a 10-pole filter in a single IF receiver. Its bandwidth is 7.5 kHz (−3 dB) for 12.5 kHz channel spacing or 15 kHz (−3 dB) for 25 kHz channel spacing. If the mixer is a diode ring then an amplifier may be inserted before the filter to remove any noise figure difficulty with the first IF amplifier device after the filter. Most receivers have dual IFs, the first is 10.7 or 21.4 MHz, the second is usually 455 kHz. The reason is that the high IF selectivity can be distributed between the two, making filters easier to implement and improving group delay flatness. The first IF amplifier following the filter is quite low gain, 15 dB or so, and is commonly a MOSFET. Its input capacitance is resonated with an inductor and a resistor added to give

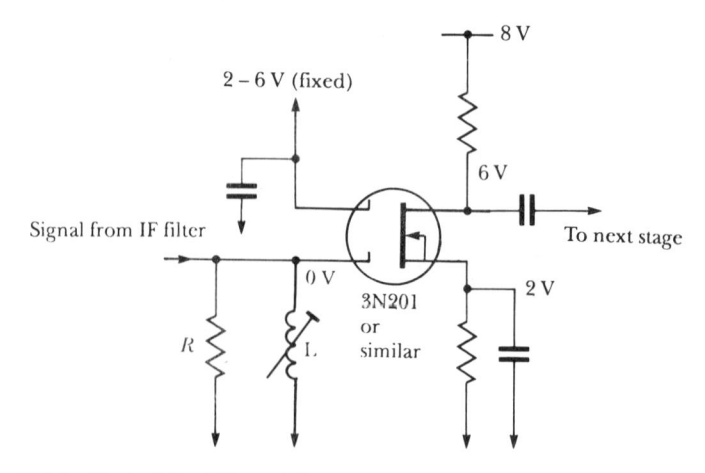

Figure 9.16 *Typical first IF amplifier using MOSFET*

Calibration, test and fault finding

a good match to the crystal IF filter regardless of signal level. A common configuration is shown in Figure 9.16 with typical DC circuit conditions.

The 'L' resonates with the FET input capacitance and this resonant circuit, in conjunction with R, matches the filter over a wide level range.

This is another device susceptible to static damage causing low gain, but possibly giving sensible DC levels. The symptom will be that the receiver is just a few decibels down on sensitivity but the quieting test will show it up. MOSFETs come in plastic 'helicopter' or metal can packages with four wires.

The second mixer down converts the first IF signal to the second, and is switched by a second local oscillator running at the difference frequency, commonly 11.155 MHz or 20.945 MHz. This mixer is usually a single device if discrete, and the oscillator usually a parallel resonant Colpitts circuit. The pair together usually take the form shown in Figure 9.17.

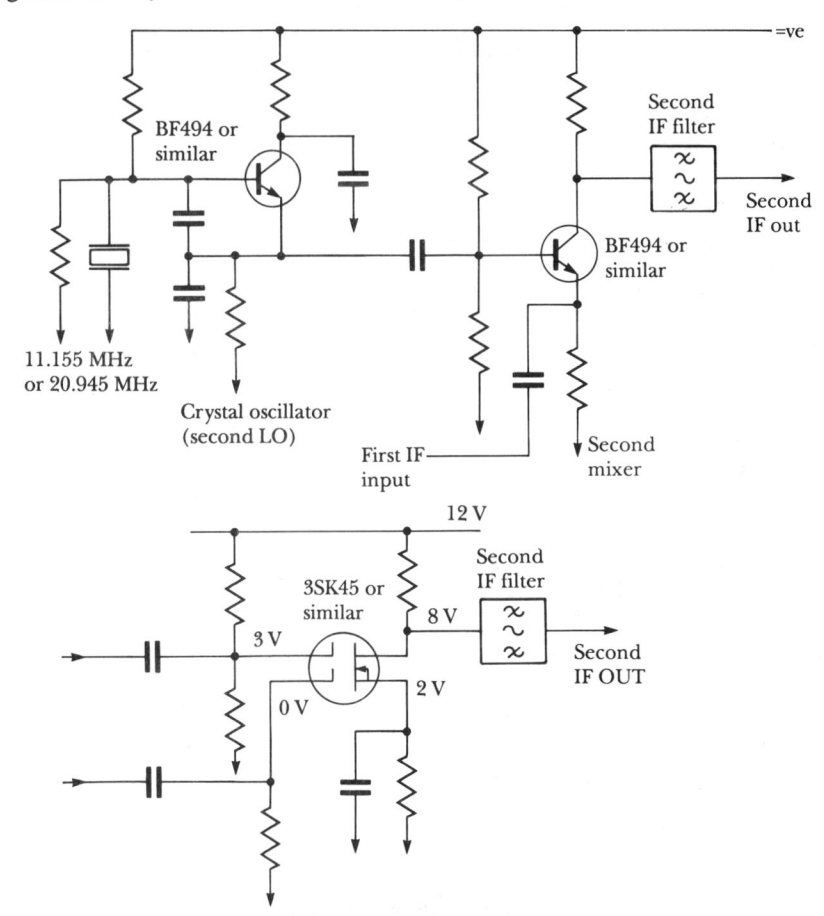

Figure 9.17 *(a) one form a second LO and mixer may take, (b) alternative second mixer using a dual gate MOSFET*

Mobile Radio Servicing Handbook

The intermodulation rejection expected of this mixer is very much less than the first mixer, since the IF filter has removed any potentially interfering signals. Also, its isolation between ports need not be high as there are high selectivity filters either side. However, because it must mix, it is run at or near zero bias if a bipolar device; a dual gate MOSFET could also be used, biased as shown.

The second IF filter is usually a ceramic device with very steep skirts; it is this filter which is responsible for the high adjacent channel rejection and typically works between impedance of roughly 300 ohms. It will have a loss of about 6 dB.

In FM sets it is after this second filter that the main chunk of IF gain is situated. The gain here is typically 100 dB which ensures limiting on front end noise and is contained in an IC with the discriminator circuits in modern receivers. A popular device is the RCA CA 3075, which contains the second mixer, local oscillator, IF amplifier, discriminator, audio pre-amplifier and squelch circuits. Another, the CA 3089, does not have squelch circuits but incorporates a delayed AGC output for reducing the front end gain under strong signal conditions. Figure 9.18 illustrates a common arrangement.

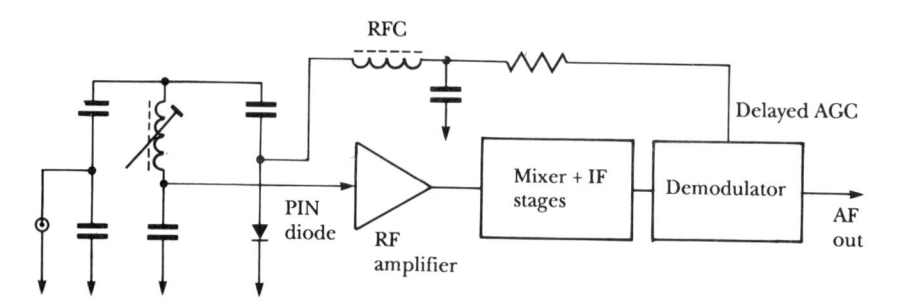

Figure 9.18 *A typical delayed AGC system*

The pin diode attenuates the RF signal ahead of the RF amplifier to reduce the receiver gain and hence increases its dynamic range under very strong signal conditions.

Older receivers may have discrete IF stages with AGC instead of limiting and a ratio FM detector; these are not considered in this book.

When this part of the circuit is suspected of having a fault, there is not much one can do without trying a new IC. However, the second local oscillator crystal will be on the board and a CRO across this will check that oscillation is present. It is worth carefully checking the DC conditions around the IC and worth checking any external discriminator coils and capacitors. An IF IC going faulty is rare and they are changed unnecessarily.

With AM receivers, the IF amplifier following the last mixer will be with an AGC loop designed to give a constant level to the detector over a wide range of

Calibration, test and fault finding

input levels. If a simple diode detector is used (the most common), a typical IF level at the detector is 1 V rms. The AGC system will be designed to reduce the gain of the stage nearest the detector first, followed by the stage before it and so on back to the IF filter. A dual gate MOSFET in this position is commonly used as an AGC element and can give about 70 dB range if the source voltage is elevated and the gate 2 voltage taken negative with respect to it. This will occur without mismatching the IF filter. Figure 9.19 illustrates a typical AM IF system.

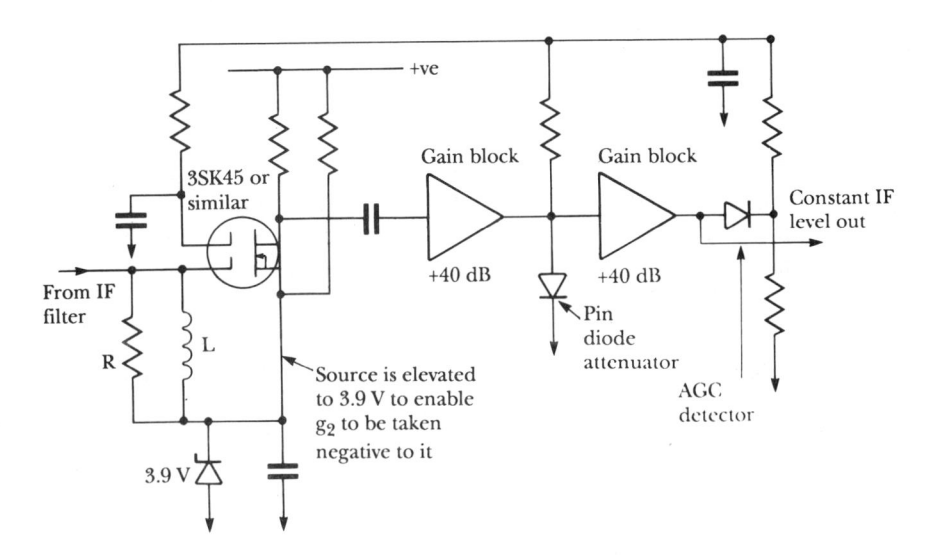

Figure 9.19 *A typical AM IF system*

1. Establish the maximum gain available by disabling the AGC using an RF millivoltmeter and signal generator on the IF frequency. As the gain is high and the final IF stage incapable of much output power, the generator output must be kept low, of the order of −100 dBm.
2. Check that each stage which is controlled responds to the control, and that the amplifier is not oscillating. Oscillating IF amplifiers happen if the set is quite old and the electrolytic capacitors have dried out and become ineffective at decoupling the IF signal. There may be a separate AGC system which operates at high signal levels to reduce the gain of the RF amplifier.

Local oscillator – as synthesizer problems are dealt with specifically elsewhere, we concentrate here on a problem between the oscillator and the first mixer, regardless of whether crystal or synthesized, and on fault finding around the crystal oscillator. The vast majority of crystal oscillators are Colpitts design which avoids the need for a tapped inductor. A Colpitts oscillator has the tuned circuit between collector and base (or drain and gate) with a capacitive tap to the emitter (or source). The crystal will be either the

Mobile Radio Servicing Handbook

resonant circuit itself in parallel mode or it will be in series with the inductor of the resonant circuit or in the feedback path to the emitter (or source), both in series mode. Parallel resonant crystals are designed to work into a certain shunt capacitance, this is usually a variable capacitor across the crystal and typically 30 pF. A series resonant crystal looks like a low value resistor typically 35 ohms in series with a DC blocking capacitor at the resonant frequency. Some common configurations are shown in Figure 9.20.

Crystal – between base and collector with tap to emitter

Tuned circuit between base and collector with tap to emitter. Crystal in series with coil

Series mode crystal equivalent at resonance

Tuned circuit between base and collector via buffer b–e junction Crystal in feedback path to emitter

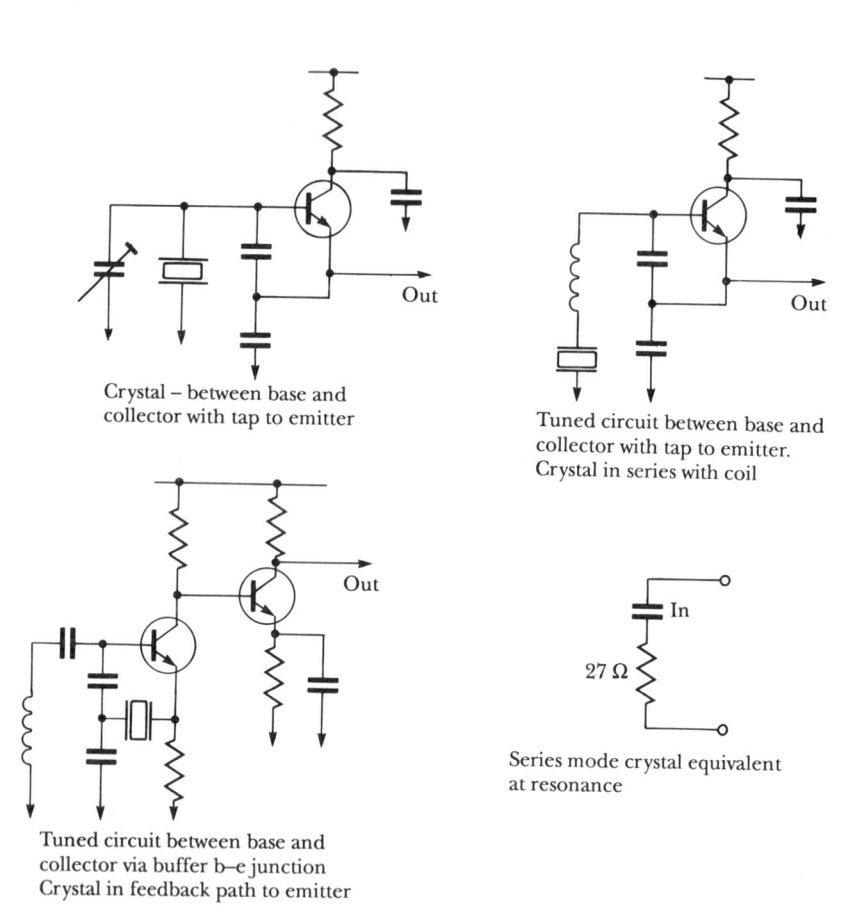

Figure 9.20 *Some crystal oscillators*

When checking the DC conditions, it is safest to remove the crystal, because under oscillating conditions the device can appear to be reverse-biased. If the crystal is suspected of low activity (i.e. low Q) it is usually checked by replacement, but if just impossible to pull onto frequency, it is possible with a series resonant circuit to replace the crystal with a low value resistor in series

Calibration, test and fault finding

with a DC blocking capacitor, whereupon the circuit could oscillate over the entire crystal range when the inductor is adjusted. Values 0f 33Ω and 1n should suffice.

Following the crystal oscillator there may be one or more multiplier stages which are designed to produce harmonics and are therefore non-linear. Two typical circuits are shown, one doubler and one tripler to give the idea.

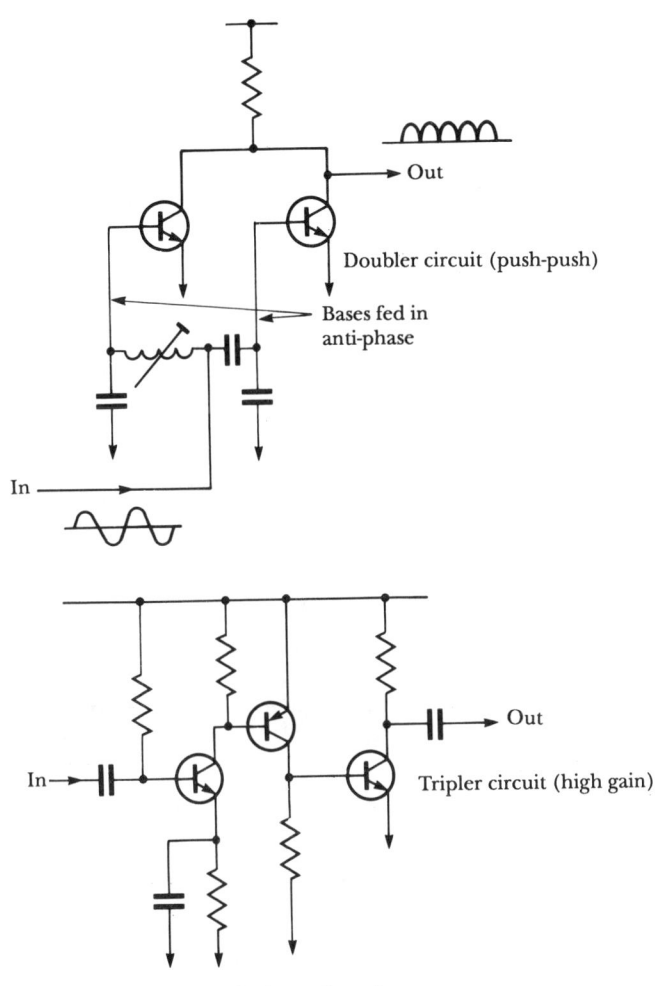

Figure 9.21 *A doubler and a tripler in configuration*

The doubler is a push–push arrangement with connected collectors to produce high second harmonic content. The resultant waveshape is similar to that from a full wave rectifier. The tripler is a high gain amplifier which limits and produces high-level odd order harmonics.

Checking DC conditions reveals most faults, again done without oscillator

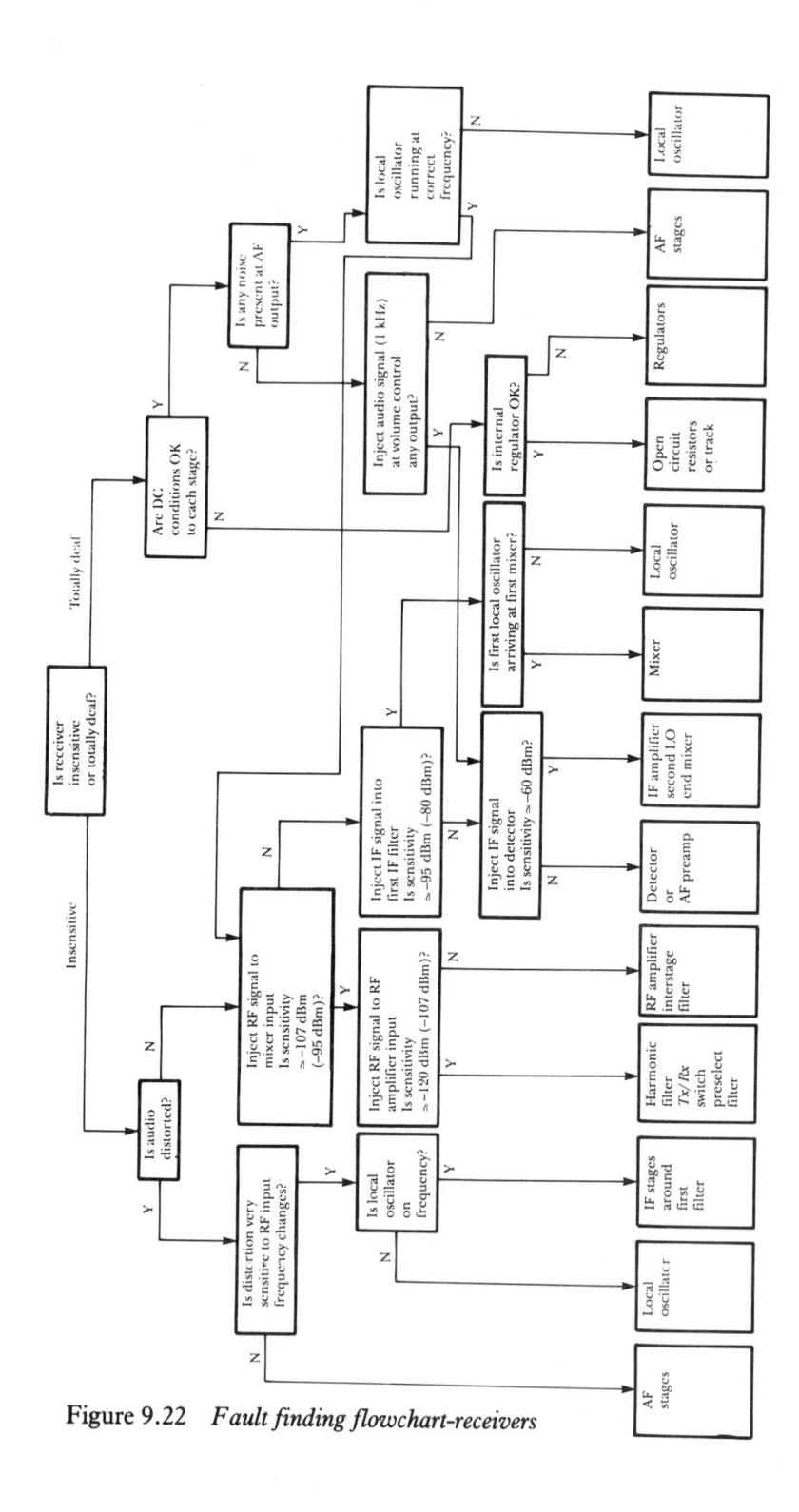

Figure 9.22 *Fault finding flowchart-receivers*

drive. The local oscillator output device may be a medium power device, especially if a high drive is required. The current that this stage draws under oscillator drive may be an indicator to its health, typically it will be 20–30 mA if driving a switching balanced mixer.

To summarize, a fault-finding flow diagram for receivers is shown in Figure 9.22. See also photographs showing the interior of a typical radio, Figures 10.1–10.5, pages 148–151.

In-service fault finding – transmitters

Introduction

This section deals with fault finding on transmitters which have once worked properly and thus assumed to have been correctly built in terms of correct components fitted. As with receivers, faults like trapped wires and dry joints may manifest themselves after a period of service, however this section concentrates on locating the fault to circuit module level and then to component level; this should be valuable in locating dry joints to a rough area, but these will not specifically be mentioned. A dry joint will normally give itself away by becoming intermittent when the circuit board is stressed mechanically. It's a case of looking for it.

Transmitters

Divide the transmitter into blocks:

1 Synthesizer or crystal oscillator
2 Modulators and multiplier chain (if crystal)
3 Exciter output stages
4 PA
5 Audio processor
6 Power supply regulators, switching logic and power control

The signal should be traceable from the source, block 1, through to the exciter output and on to the PA. It's a case of regarding the transmitter as a chain of blocks and finding where the signal is lost, or the block which refuses to tune properly. To help locate a faulty block, listed below are the blocks and typical power levels expected from them:

Measurement point	Level (50 ohms)
Crystal oscillator output/output from synthesizer VCO	+10 dBm
Multiplier chain output	+20 dBm
Exciter output to PA	300 mW to 5 W

Checking each stage with a diode-probe voltmeter is a useful indicator to a stage which will not tune properly.

If RF is present but no modulation is possible, then the fault lies in the AF block.

Some of the blocks are now considered in more detail.

1 *Oscillators*

The same methods apply to transmitters as to receivers. Crystal fault finding is covered in detail in Chapter 7. If synthesized, the section on synthesizer fault finding gives useful information, and the reader is referred to these sections accordingly.

2 *The exciter*

a) Crystalled sets: Following the crystal oscillator will be one or more stages where phase modulation is performed. The number of stages depends upon the multiplication number up to the final RF frequency and ranges from 1 or 2 UHF sets to 4 or more in sets operating in VHF low band with a multiplication number of 3.

A common phase modulator configuration was shown in Figure 9.8. The stage is arranged as a phase splitter, and most phase modulators work using a PIN diode element as a variable resistor against a reactive element, or sometimes as part of a resonant circuit. Phase modulators are sensitive to RF input level: if too low they give too low modulation. If too high they distort giving harmonics of the modulation tone. DC conditions should be measured around this circuit in absence of oscillator drive.

Following the modulator stages there will be one or more multiplier stages and the same comments apply as to those associated with receiver local oscillator multipliers. Fig 9.21 gives typical circuits.

b) Synthesized and crystalled sets: The modulated RF signal at the final frequency is amplified by one or two further stages operating at or near Class C up to a level suitable to drive the PA stage A which works in class C. It is the exciter output stages which are generally used as the variable drive element(s) to maintain a constant level from the PA output within a power control loop. An input here also arrives from the synthesizer to inhibit the PA when it is out of lock. This means that their current is varied by a series pass device controlled by a hot-carrier sense diode which samples the PA output. A typical arrangement is shown in Figure 9.23.

The series pass device is fed from a DC amplifier and potentiometer arrangement fed from the sense diode to maintain the PA output power at a level set by the potentiometer. It is important that the PA be tested with no signal fed back via this diode, i.e. open loop giving maximum available power. It is essential to check DC conditions around a circuit such as this; the series pass device is quite highly stressed and if the exciter output stage draws excessive current through it due to a fault in the matching circuits, it may well go short-circuit, thus disabling the power control.

Calibration, test and fault finding

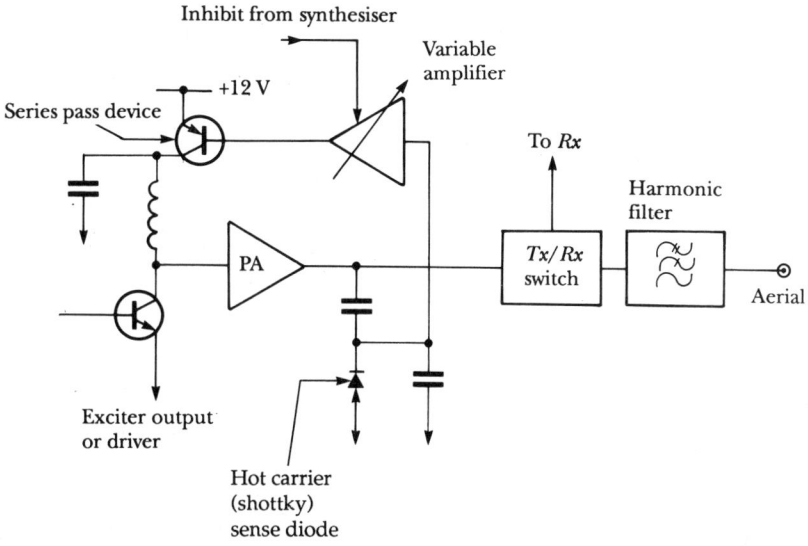

Figure 9.23 *A power sense and control configuration*

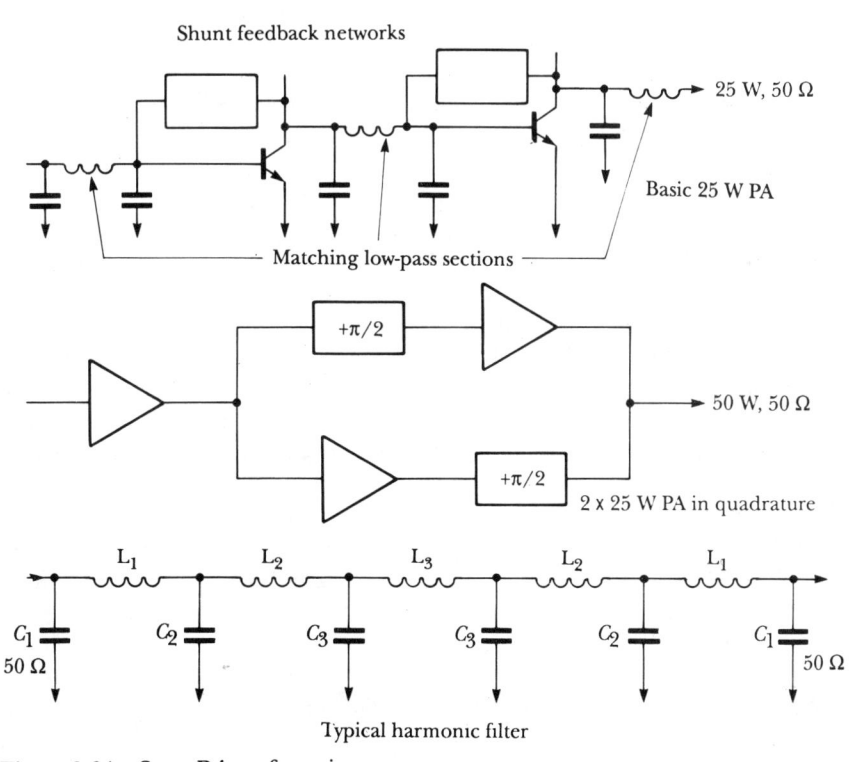

Figure 9.24 *Some PA configurations*

139

3 The PA Stage

Usually two stages are involved; driver and final. PAs rated to 25 W usually only have a single final device, above this two may be used. The operation is Class C with low-pass filter matching sections between the stages and between the final and the harmonic filter, which is again low-pass but usually symmetrical. Some typical PA arrangements are illustrated in Figure 9.24.

If a PA does not perform and the drive to it is confirmed as OK, then the faulty stage can usually be determined by the amount of current drawn. If under 1 A, then the driver is not conducting. If 1 to 1.5 A, then the driver is probably alright and the final is suspect. If 3–6 A, the final is alright and the problem lies between it and the antenna connector, i.e. in the harmonic filter or antenna switching circuitry. PA transistors are usually stud or bolt-down devices with ceramic bodies some 10–15 mm diameter, and contain beryllium oxide discs which bond the chip to the stud. If this disc is broken, the device heats very rapidly with subsequent fast power drop-off. This is a common cause of PA device failure. Replacing these devices commonly involves removal of several chip or cased mica capacitors and takes about thirty minutes and they are quite expensive (some cost over £30), it is therefore important to diagnose correctly as they cannot be re-used.

The harmonic filter is usually symmetrical and if suspected, a power meter can be placed at the PA side of it to check its loss, which should be <0.1 dB.

The transmit/receive switch is either a solid state pin-diode switch or (less commonly now) a coax relay. If a relay it will be energized by the PTT switch, and a PIN diode switch usually will be activated from individual switched receive and transmit regulated supplies generated within the radio. There will be one diode in series with the transmit path from the PA, and another in shunt or in series with the RF path to the receiver, to prevent RF from the PA destroying the receiver front-end on radios where the transmit and receive frequencies are equal.

In AM transmitters the PA current is carried through a transformer winding to enable amplitude modulation. The power control loop, if present, must be slow enough not to clean off the modulation. It is possible to use a series pass device to modulate PA current, then the same applies as to the exciter section above in terms of fault finding. This method is somewhat unreliable and less efficient than using a transformer. There will also be a means of switching the loudspeaker off during transmit, usually accomplished with a relay. The modulation transformer itself has been known to develop shorted turns etc., which causes grief to the audio output stage, particularly if the PA has drawn excessive current and overheated it.

4 Audio circuits

The path from the microphone terminals or line input etc. to the modulator stage, passes through some sort of audio compressor stage; one typical configuration is illustrated in Figure 9.25.

The same basic method is used in most compressors. The output is rectified

Calibration, test and fault finding

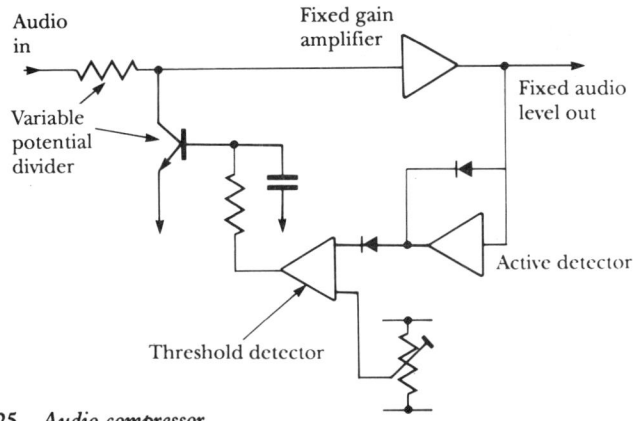

Figure 9.25 *Audio compressor*

and compared with a threshold voltage, and when exceeded by the rectifier output the comparator signals to a device which forms one half of a potential divider at the input to reduce the level accordingly. Thus the output level remains constant over a wide range of input signals and remains substantially distortion free. A typical dynamic range is 50 dB at 1 kHz.

Fault finding around a compressor stage is more difficult because it is a closed loop, and most faults occur in the detector and comparator stages. If the feedback signal to the active potential divider device is temporarily disconnected then the detector and comparator can be checked individually with a CRO and voltmeter, with an AF generator on the input.

There will also be a pre-emphasis stage, a limiter and possibly an integrator stage on FM sets. These circuits are normally arranged around multiple op-amp chips and some common configurations are illustrated in Figure 9.26.

All op-amps used in these configurations have an average DC output midway between their supply rails, and this is the first thing to check when fault finding. An op-amp whose output terminal is hugging either supply rail is either suspect or is having its biasing upset. A CRO on the output of each stage in turn will help identify the faulty stage.

Logic and timing circuits

The receiver/transmitter changeover switching is commonly done with CMOS gates switching medium power devices to apply power to the receiver or transmitter. It is usual to leave the Class C PA and exciter output stages with power applied permanently since these will not draw any current until driven. Thus the current drawn by the switched stages is quite small, under 1 A.

A maximum transmit timer is usually incorporated in case the PTT is left depressed to avoid overheating; mobile transmitters are not continuously

Mobile Radio Servicing Handbook

Pre-emphasis

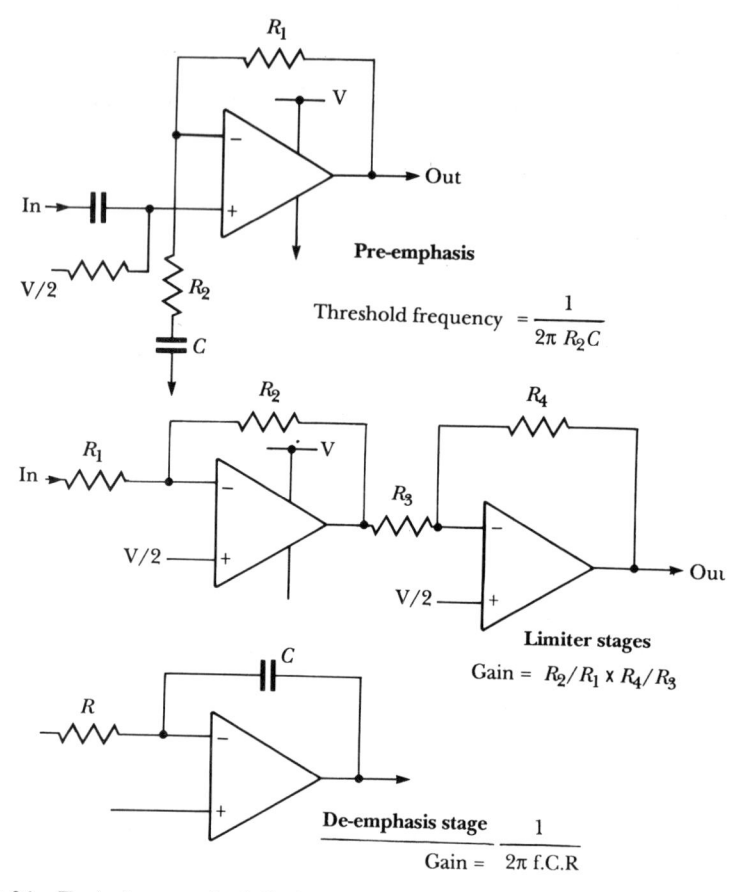

Threshold frequency $= \dfrac{1}{2\pi R_2 C}$

Limiter stages

Gain $= R_2/R_1 \times R_4/R_3$

De-emphasis stage

Gain $= \dfrac{1}{2\pi \text{ f.C.R}}$

Figure 9.26 *Typical pre-emphasis limiter and integrator stages*

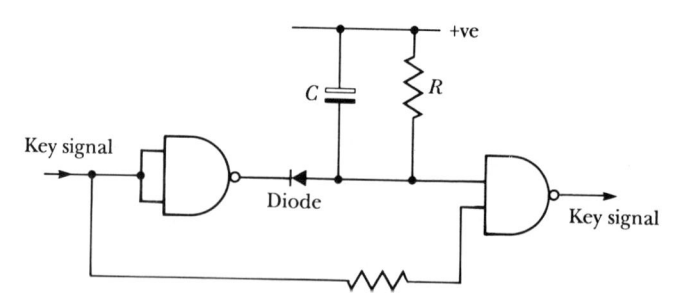

Figure 9.27 *Transmit timer*

rated. A typical maximum transmit time is 30 seconds to 1 minute. One arrangement used is illustrated in Figure 9.27.

CMOS gates used in transceivers are prone to static damage and if faulty can

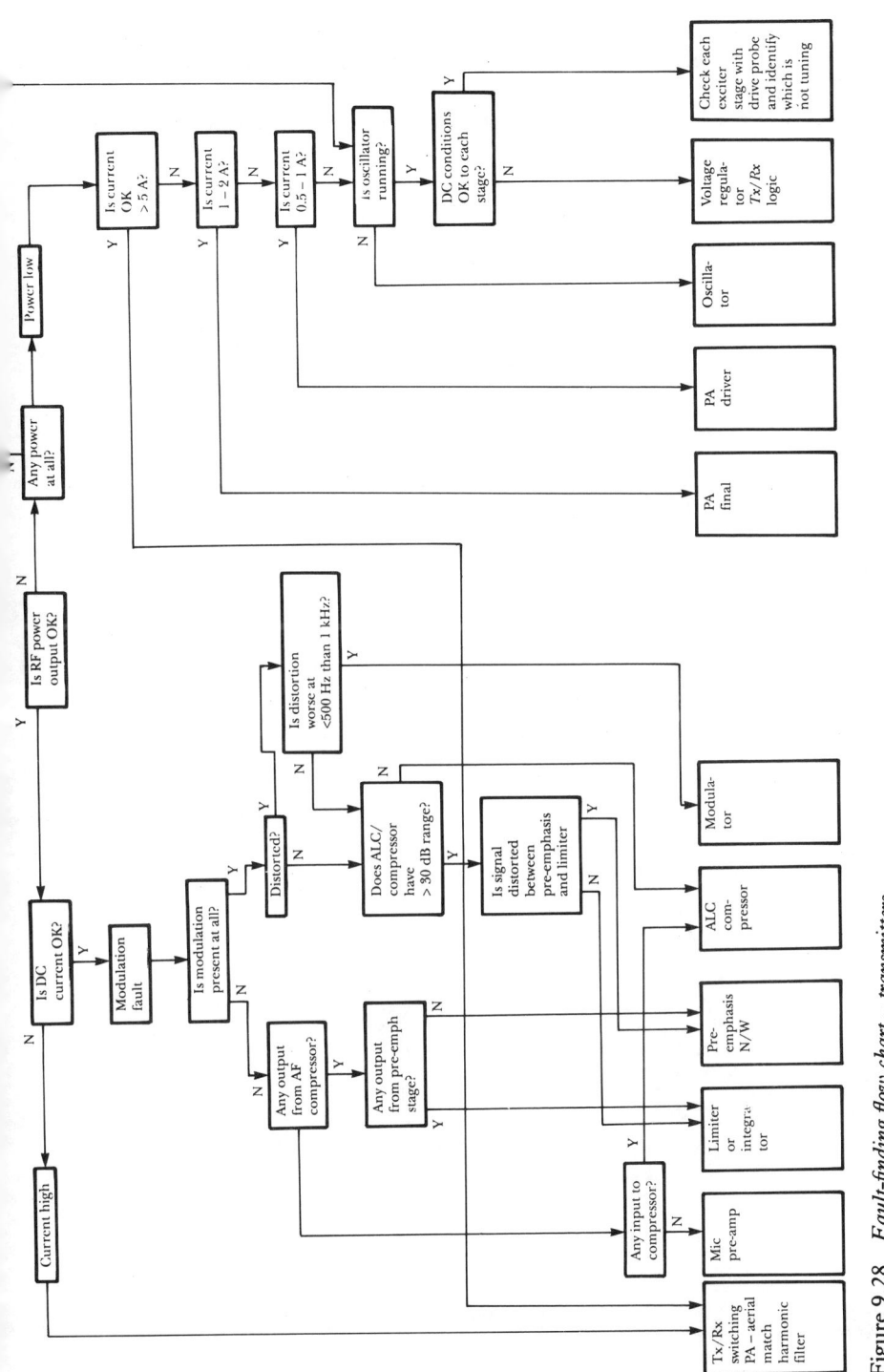

Figure 9.28 *Fault-finding flow chart – transmitters*

result in no receive, no transmit, or both together with subsequent possible damage to the front end.

The timer is usually based on a CR discharge characteristic with a diode to ensure rapid Rx to Tx switching with fast recharge of the capacitor.

To summarize, a fault-finding flow diagram for transmitters is shown in Figure 9.28. See also photographs showing the interior of a typical radio, Figures 10.1–10.5, pages 148–151.

10 Component identification and handling

Classification of components

When confronted with an unfamiliar radio of dubious vintage it is useful to identify each component by its appearance before attempting to classify it electronically. Basically there are four groups of components:

Passive components

These consist of resistors, capacitors and inductors, and according to their value, tolerance and power rating, come in a wide variety of packages. By far the most numerous will be simple resistors, then capacitors, then small inductors, followed perhaps by larger power resistors, and finally large electrolytic capacitors and iron-cored inductors associated with the power supply decoupling. Variable resistors (potentiometers) may either be skeleton preset types mounted on the PCB (for power control, modulation depth etc.), or large rotary controls (volume, squelch, sensitivity etc.) with spindles. As a rule, those potentiometers which, if adjusted, could increase the interfering potential of the radio will be pre-set types inaccessible to the normal user.

All capacitors have two connections; variable types may have more legs, but in that case two or more of the legs will be connected together. All fixed resistors have two legs; most variable resistors have three, i.e. one connected to either end of the track and a third connected to the wiper. Most small inductors used in tuned circuits have more than two legs, the extra legs being connected to taps along the coil length. However, only two of the legs will be connected into circuit, and different pairs of legs will be connected according to where the radio is tuned in its band. These coils are normally wound on a vertical former and will probably have an adjustable ferrite dust core and a screening can. Inductors used as chokes will only have two legs and sometimes they have a stripey body much like a resistor, and are mostly mounted horizontally.

Chip components are harder to identify. Currently, chip capacitors are about 3 mm square up to 5 mm square with two opposite sides having metal deposited to aid soldering. They are generally not marked with the value and are really only checked by insertion in a measuring bridge or by substitution with a known value. Chip resistors are also about 3 mm square with two opposite metal sides and current practice is to mark on their value using three digits such as 331, which indicates 33 multiplied 10 to the power 1, i.e. 330 ohms. By the same reasoning, $105 = 10 \times 10^5 = 1$ Megohm.

Discrete semiconductors

Discrete semiconductors include transistors, temperature dependent and non-linear resistors, crystals, and diodes.

Transistors come in two basic types, bipolar and field-effect, and these are each subdivided into two polarities and a wide range of shapes and sizes for each. However, they are fairly easily identified. All bipolar transistors have three legs connected to their active parts, plus a possible fourth which will be connected to its metal case or to another of the legs. RF devices with four legs are either the 'pill' type with two opposite legs connected to the emitter, or are the 'top hat' metal can type with the fourth leg connected both to the can and to an internal screen.

Bipolar transistors come in two basic polarities, PNP and NPN and their applications are discussed later in the chapter. It is important to look up the device in a data book to identify the polarity.

Field effect transistors (FETs) come in two polarities, N-channel and P-channel, and each can either be depletion-mode or enhancement mode. On top of that, JFETs use a junction like that of a bipolar transistor for the gate interference to the channel, but a MOS FET has a fully insulated gate (they are sometimes known as IGFETs). The package will be plastic or metal, and power FETs are now available in a TO–220 type package, so it is essential to check their part number in a data book as they are often indistinguishable in appearance from bipolar transistors.

Temperature dependent and non-linear resistors fall into the discrete semiconductor category as these are bulk semiconductor devices with the doping designed to obtain the desired characteristics. Temperature dependent resistors come in two types, positive temperature coefficient (PTC) and negative temperature coefficient (NTC). They are made in various power ratings and in appearance they are usually like a medicine pill, about 5–10 mm in diameter and 1–2 mm thick with wires attached to either flat surface. They are painted with horizontal coloured stripes in the normal resistor colour code and the body colour usually gives the type. The body colour will be referred to in the appropriate data books, but a starting point is the Mullard (Philips) data book as over 50 per cent of PTC and NTC resistors, in the author's experience, are made by this Company. Non-linear resistors alter their value according to the current flow and are used mostly as surge arrestors but are relatively slow acting. They are rarely found in modern equipments and are not discussed further.

Crystals, although quartz, have been included under this heading. Crystals will be in a metal can with two legs which plug into a holder, or two wires which solder directly into a PCB. The most popular packaging now is a cold-weld HC25 case which measures about 15 mm high by 12 mm wide by 2 mm thick with rounded ends. The operating frequency is usually printed on one side. It is likely that the receiver will have one or more crystal filter blocks containing several crystal elements. This will be a metal can, perhaps 20 mm long by about 13 mm square with four legs; input, output and two for ground close to the input and output.

Ceramic resonators are a common sight in receivers, used both as a multi-element filter at a low intermediate frequency (IF) or as the resonant circuit element in the FM demodulator. In the first case, it is a plastic square package with four legs; input, output and two grounds, near the input and output. A plastic square package is used for the demodulator resonant circuit but this is a single element and as such will only have two legs. Each will carry a shortform code moulded or printed onto the plastic and it is a case of hunting through the data books again to find further information.

Diodes are the most common discrete semiconductor devices used in radio and all have two connections. Signal diodes are all tubular and mostly glass packaged, some medium power diodes are plastic packaged. High-power diodes are stud-mount and bolted onto the chassis. Signal diodes come in various forms – popular are the schottky diodes, zener diodes, PIN diodes, ordinary fast diodes, etc., and all these types are discussed with their application later in the chapter.

Integrated semiconductors

An integrated circuit in most cases gives itself away by having more than four legs and a plastic, ceramic or (more rarely) metal body, but there are exceptions. Look out for three-terminal regulators which look just like plastic TO–110 or metal TO–3 power transistors, and also for darlington devices which have two transistors in one package and three legs looking just like a normal bipolar transistor. Anything unusual should be checked in a data book before decisions about replacing it are taken.

Special ironmongery

Modern specification on radio equipment has led to the development of cavity resonators at UHF and to special screens placed strategically around the radio to reduce undesirable effects. A cavity resonator will normally have a tuning screw maybe 3–5 mm in diameter protruding from the top and will be a fabricated or cast series of rectangular cavities, probably containing helical resonator coils, and be bolted firmly down onto the PCB. Screens may be

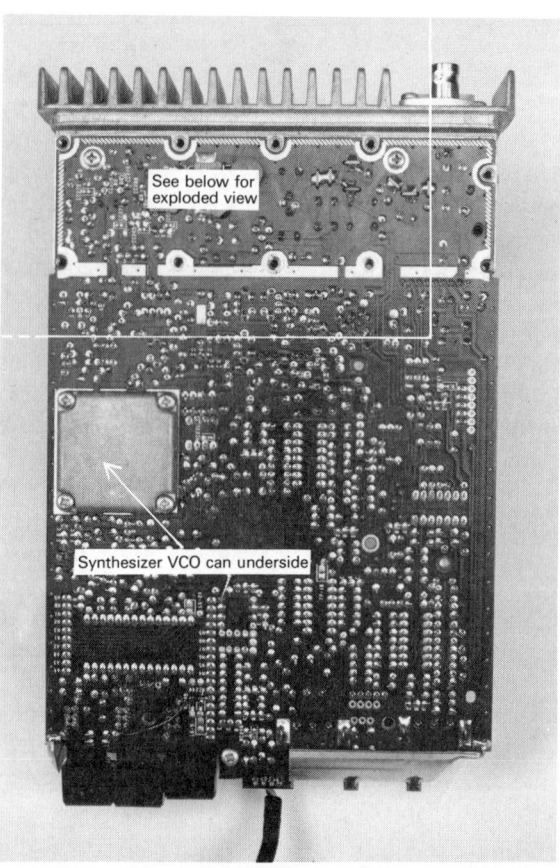

Figure 10.1 *Underside view of typical transceiver (PA shield removed)*

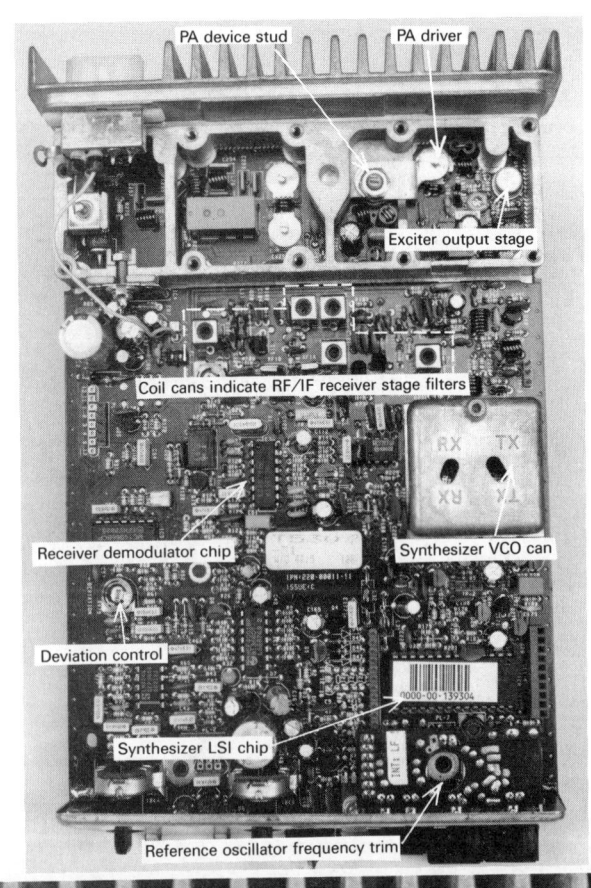

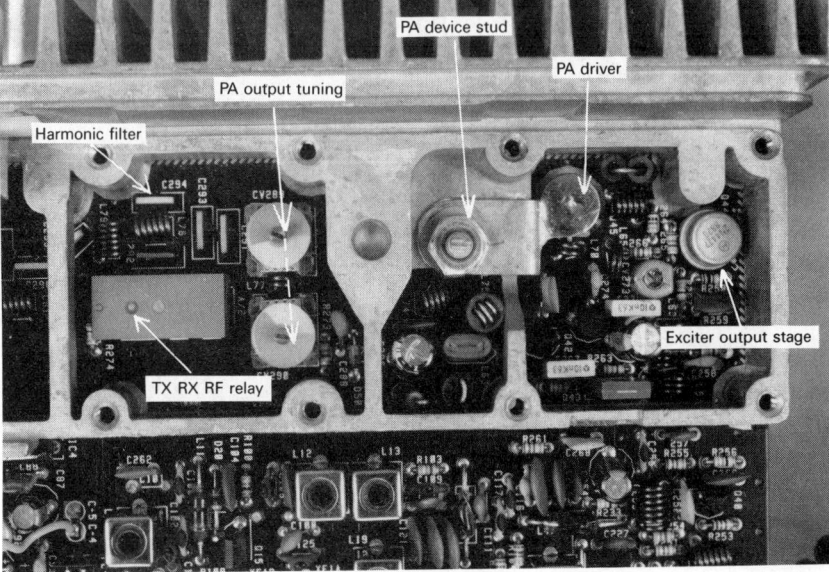

Component identification and handling

Mobile Radio Servicing Handbook

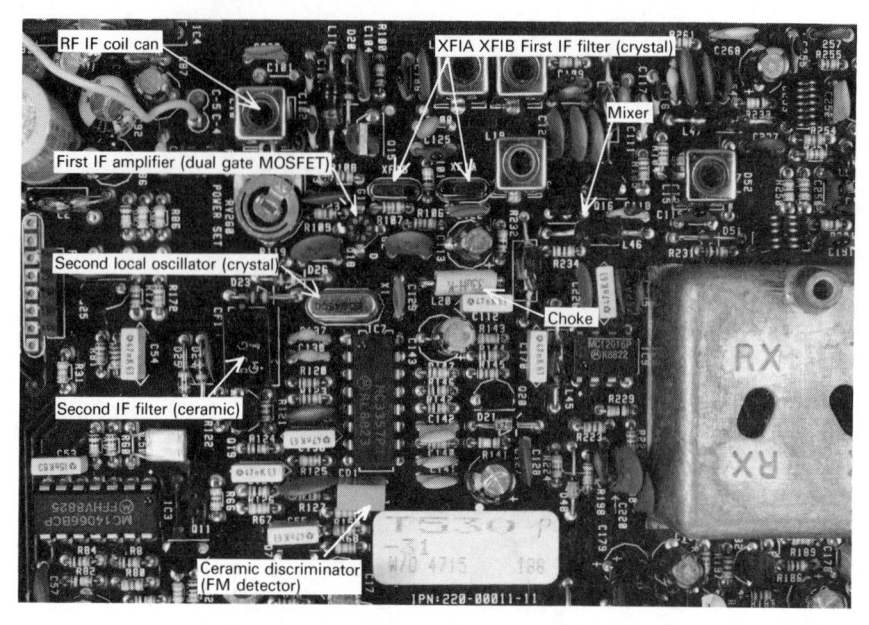

Figure 10.2 *topside view of typical mobile transceiver (PA shield removed)*

Figure 10.3 *Close up of receiver section of typical transceiver*

Component identification and handling

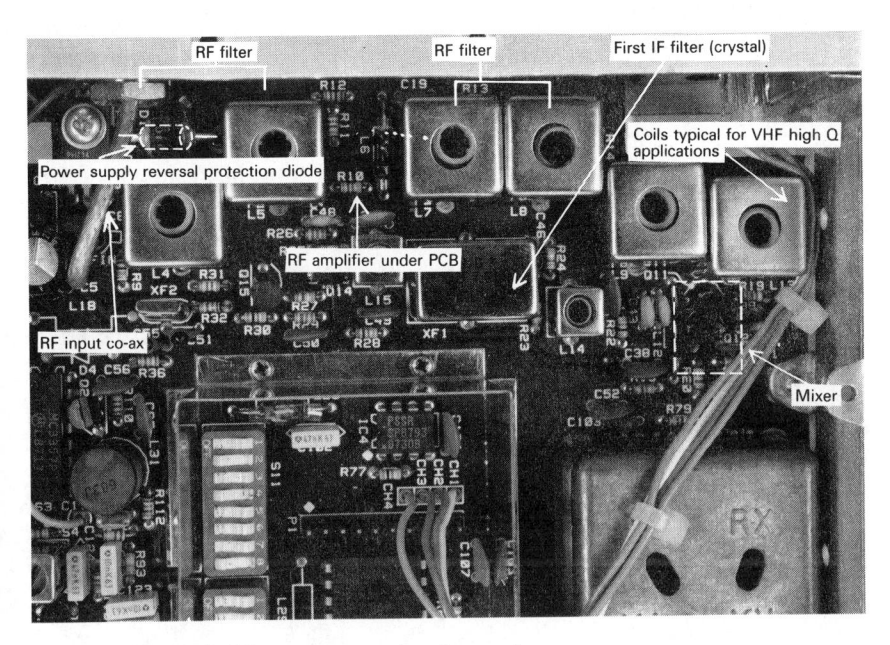

Figure 10.4 *Typical base station receiver front end*

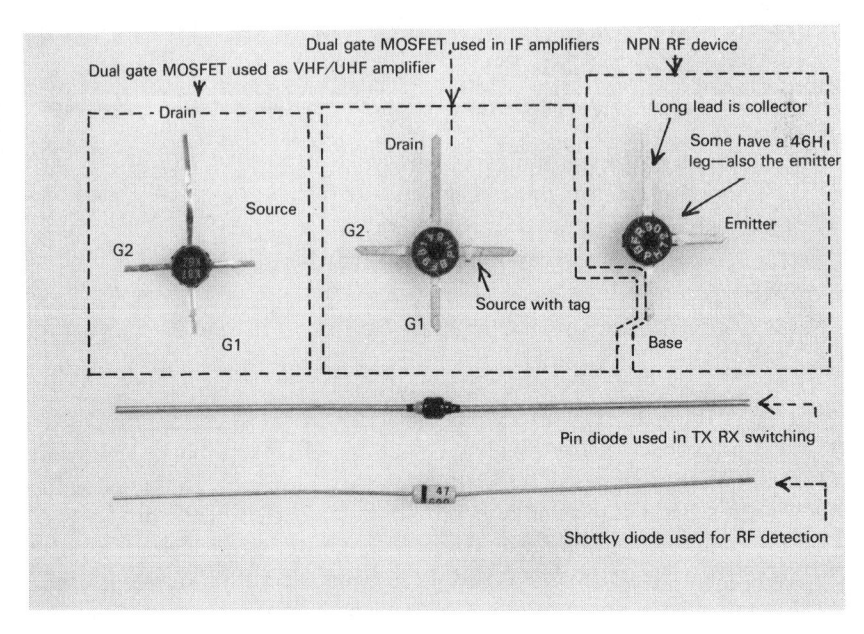

Figure 10.5 *Some semiconductors*

present on one or both sides of the PCB and will vary from being simple short 'walls' to complicated low profile boxes enclosing a number of components. It hardly needs mentioning that if any screens are disturbed during servicing they should be carefully replaced.

Figures 10.1–5 are photographs of the interior of a typical radio with the subsections marked and the various components identified.

Progress in manufacturing techniques has enabled sophisticated devices to be made cheaply enough for use with mobile radio. Integrated circuits are currently found in all audio stages, IF and detector stages and switching and timing circuits. Large scale integrated circuits are now finding uses in synthesizers and microprocessors used in controlling radios and for signal processing. Work is continuing on integrated circuits for RF into the UHF region. Some transmitter PAs consist of encapsulated modules. Defining some possible circuit positions and some handling procedures will be of benefit to the service technician who perhaps is working with less than ideal documentation.

Discrete semiconductor device applications

Discrete devices are found, at present, in the following stages of a typical mobile radio transceiver:

1 RF amplifier
2 Mixer
3 First IF amplifier
4 Any additional IF amplification required above that offered by ICs used
5 Oscillators
6 RF amplifiers in the transmitter chain
7 High Current regulators or switches
8 Modulator AF stages in AM transmitters
9 Simple low-signal amplification or switching where economical
10 Diodes

1 RF amplifier
This is nearly always bipolar NPN if the radio is UHF (470–520 MHz), Band III (174–225 MHz), or VHF highband (138–174 MHz). Other device types are used for frequencies below highband such as JFETs, bipolar PNP and occasionally MOS dual-gate FETs, although these are not recommended for this application because of their susceptibility to damage by static discharges.

The most common configuration is common-emitter if the device is bipolar NPN, to achieve the desired gain and noise figure. If a PNP device is used, chances are it will be common-base. Common-base stages have the advantage of zero reverse transmission coefficient and so will not put any local oscillator

Component identification and handling

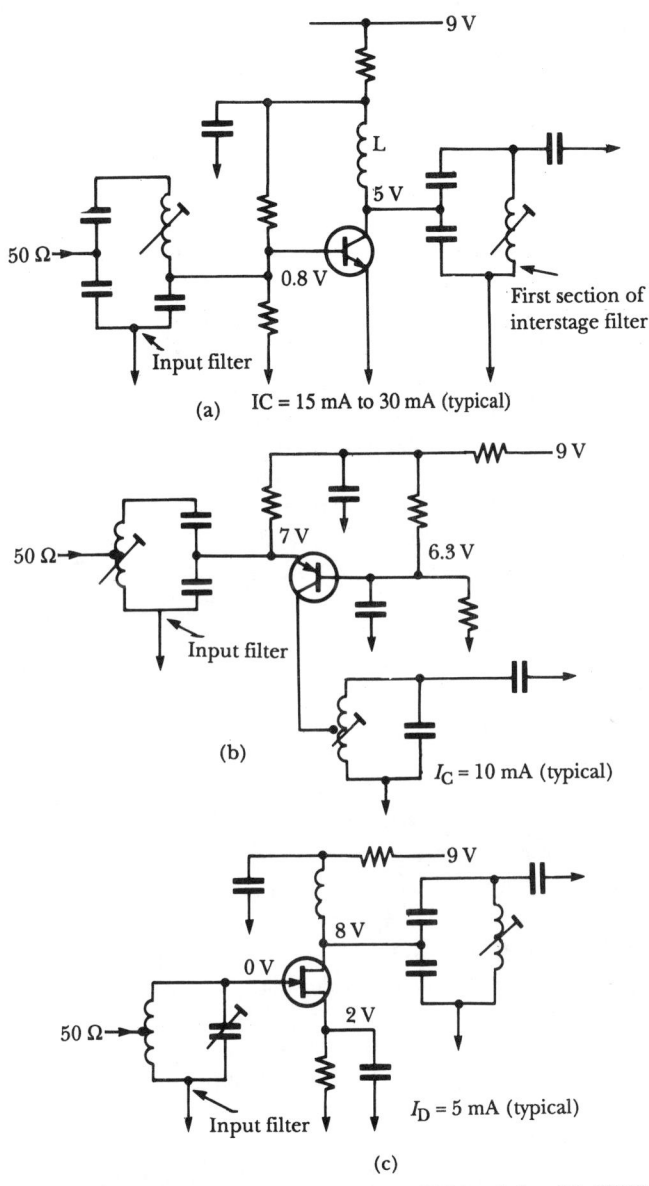

Figure 10.6(a) *NPN device in common-emitter, use: highband, band 3, UHF, (b) PNP device in common-base, use: lowband, highband, VHF, (c) JFET in common-source, use: lowband, highband, band 3*

signal onto the antenna. JFETs connected in common-gate do not share this advantage. A JFET in common-source has an advantage in that its input impedance is an order of magnitude higher than a bipolar device, which means less load on the receiver input tuned circuit, higher Q, and hence less loss and lower overall receiver noise figure. However, the best JFETs do not have as low a noise figure as a good bipolar device which offsets this advantage.

JFETs need fewer components around them to bias them, and this is a real advantage in manufacture. Economy in components is the deciding factor when choosing an RF amplifier device when there is a choice of devices all giving similar performance. For this reason, a PNP device may be used. In older designs of AM sets, dual gate MOSFETs were most economical in components when an AGC RF amplifier was required.

A few typical biasing arrangements, voltages and currents are shown as Figure 10.6. Figure 10.6a shows an NPN device in common-emitter. Note that capacitive matching is used to economize on components: the capacitors also act as DC blocks. L is a choke presenting a high impedance to the collector so that the amplifier load is almost entirely the interstage filter.

Figure 10.6b shows a PNP device in common-base. Note the economy in components when the collector is at ground potential.

Figure 10.6c shows a JFET in common-source. Note that the gate is at ground potential, and the gate-source capacitance is tuned out by the input filter resonant circuit.

2 Mixer

The first mixer is the most critical stage in the receiver when considering its ability to handle high level interfering signals. It is commonly a balanced configuration using bipolars or JFETs, and may use medium power devices in base station receivers conducting 50 mA or more. Again, JFETs use fewer components to set their DC operating conditions, but if currents of the order of 20 mA or more are required for adequate third order intercept performance, bipolars are necessary since it is difficult and expensive to source JFETs to handle this magnitude of current at VHF. UHF mixers are not normally balanced active devices. Single active devices can be used, but for top performance it is normal to find a balanced shottky diode-ring mixer.

A few typical biasing and local oscillator drive arrangements, together with DC voltages and currents are given in Fig. 10.7. Figure 10.7a shows a bipolar balanced mixer capable of +30 dBm third order intercept point.

Figure 10.7b shows the equivalent mixer using JFET's. In both these types, the transistors are biased such that they are 'just' off, and the local oscillator (LO) switches them.

Figure 10.7c shows a single JFET configured with helical filters for UHF use. Note the simplicity of connection and the beauty of a gate electrode at ground potential.

3 First IF amplifier

The first IF amplifier has two main constraints; it must provide a reasonable match to the first IF filter output over a wide range of signal level, and it must

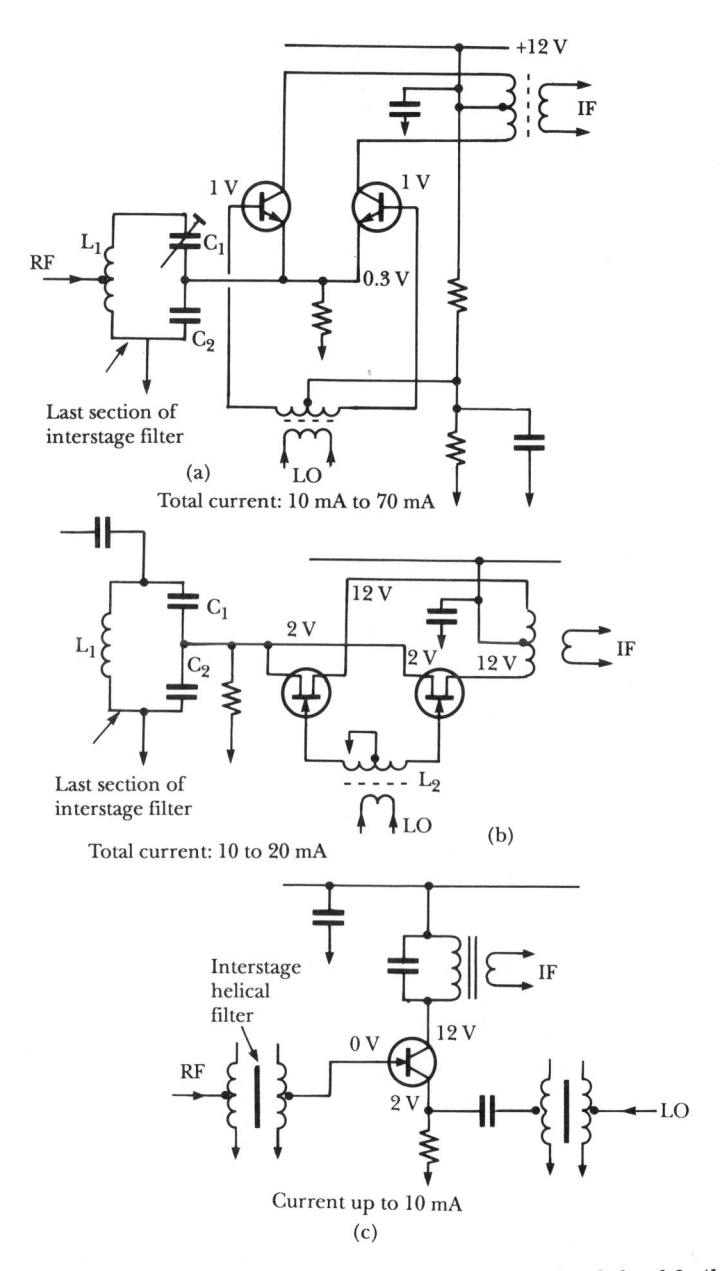

Figure 10.7 (a) *Bipolar balanced mixer, use: lowband, highband, band 3*, (b) *JFET balanced mixer, use: lowband, highband, band 3*, (c) *JFET, use: UHF*

have a reasonably low noise figure (of the order of 6 dB). There may be an additional requirement of AGC application to the stage, particularly in AM receivers.

155

Mobile Radio Servicing Handbook

One device which meets these requirements admirably is the dual-gate MOSFET. It is too well buffered from the antenna to be troubled by static discharges, and requires few components to set up its DC operating conditions. Its gain will not be high, roughly 15 dB, and this is satisfactory for the majority of dual conversion receivers as the first IF amplifier gain before the second mixer. The frequency of operation is commonly 10.7 or 21.4 MHz which, for cost and space reasons, rules out an integrated circuit amplifier. AGC can, if required, be applied to the second gate. A typical arrangement is shown in Figure 10.8.

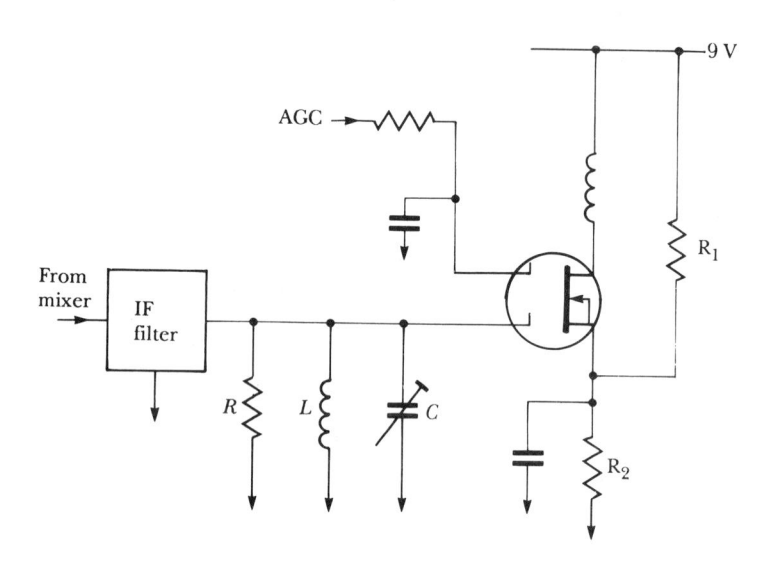

Figure 10.8 *First IF amplifier with AGC using dual-gate MOSFET*

Note that the source is held positive with respect to ground by R1, R2, even when the device is conducting no current. This is so that the gate 2 voltage can become negative with respect to the source to obtain a large AGC dynamic range of perhaps 70 dB. R, L and C in the figure provide the correct filter termination.

4 Additional IF amplification

Additional IF amplification is sometimes necessary when the minimum specified (worst case) gain offered by the IC amplifier at the second IF is deemed insufficient for satisfactory limiting characteristics. Too low an IF gain will make the receiver susceptible to ignition noise and other amplitude related interference. NPN bipolar or JFET devices are used according to amount of gain required and number of components allowable.

Component identification and handling

5 Oscillators

Devices used in oscillators are nearly always discrete and nearly always NPN bipolar or JFET, at any frequency up to UHF. Low noise is essential for good signal: noise and receiver blocking performance and IC oscillators are bulky and more noisy. A few examples of device circuit position and DC voltages and currents are given in Figure 10.9, both for LC and crystal oscillators. Figure 10.9a shows a bipolar Colpitts design using a series-mode crystal. R can be quite small (a few hundred ohms), which suppresses spurious crystal modes.

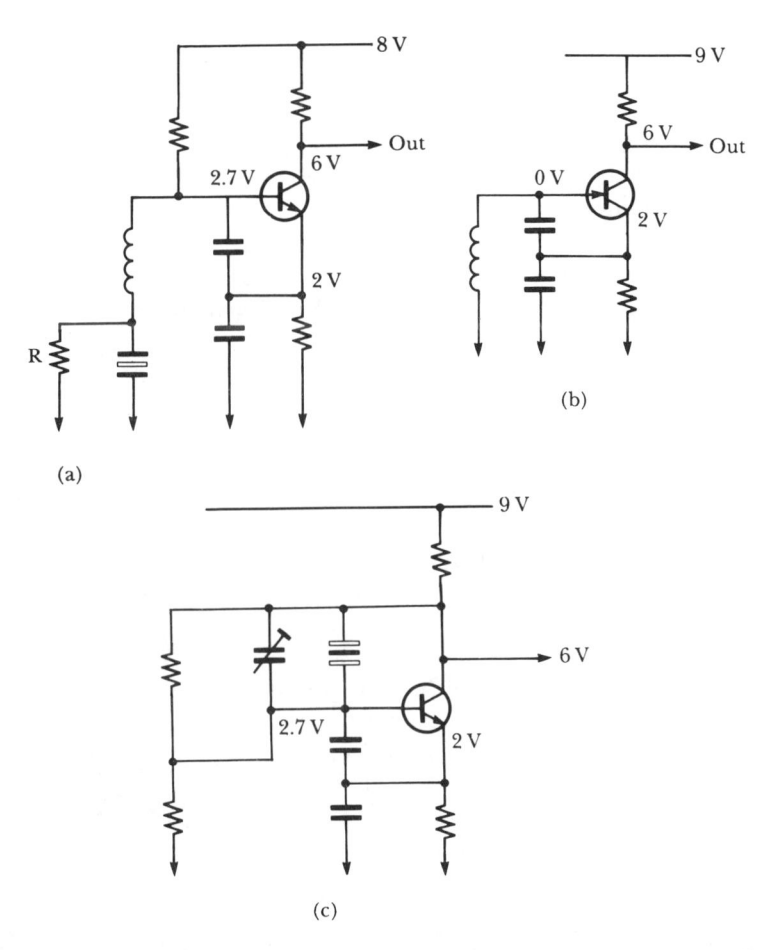

Figure 10.9 (a) *Bipolar Colpitts with series resonant crystal, use: to above 60 MHz*, (b) *JFET Colpitts using L-C to UHF*, (c) *bipolar Colpitts using parallel resonant crystal, to 20 MHz*

Figure 10.9b shows a JFET Colpitts oscillator which will run at any frequency to UHF. Note that since the gate voltage is zero, direct connection of the inductor between the gate and the ground is possible. Thus, a length of semi-rigid coax or similar can be used.

Figure 10.9c shows a bipolar device with a parallel resonant circuit also in Colpitts configuration.

6 *RF amplifiers in the transmitter chain*

RF amplifiers in the transmitter chain are discrete for at least some of the way, using devices of gradually increasing power dissipation capability towards the PA stages which may be a module or two more discrete devices of even higher power rating. This applies to crystalled as well as synthesized transmitters. The RF stages under discussion are almost always bipolar NPN, and when operating in Class C, require very few components to set the DC operating conditions. Most of the components will be to achieve optimum impedance matching over the operating frequency range.

In crystalled sets, devices in multipliers may be teamed in pairs (push–pull or parallel) to improve efficiency.

7 *High current regulators*

High current regulators handling say 50–100 mA are employed to drop the nominal input DC voltage which may be 13.8 V to, for example, 5 V for microprocessors or 8 V as a regulated supply line for the small RF signal circuits. A three terminal TO–220 package regulator may be used for the purpose, with its tab bolted to a heatsink or large copper area on the PCB. Quite common also are discrete regulators with inbuilt current limit. The advantage with discrete regulators is that their drop-out voltage is lower, i.e. the voltage they must have across themselves. A three terminal integrated regulator requires 2 V or so, which when added to an output voltage of 8 V gives a minimum input voltage of 10 V. Add to this 1 V for the forward drop of a protection diode, and suddenly the minimum voltage which the radio can operate from is 11 V. Most radios are specified to operate from 12 V ±10 per cent which is then a problem. So three terminal regulators are used for 5 V production, and discrete regulators for 8 or 9 V, since their drop-out voltage can be as low as 0.5 V if they use a PNP series pass device.

High current switches are used to apply current to the transmitter RF circuits and remove it from the receive circuits during PTT. One is also commonly used to switch the whole radio on and off if the set is mounted remotely from the front panel, so that the connecting cable (which might be quite long) does not have to carry the supply current. PNP devices are the most popular for this for two reasons. The first is that their saturation voltage is lower than for NPN devices, and the second is that their bases are taken negative with regard to their emitters for switch-on. With an NPN device it is necessary to apply a current to the base from a voltage supply higher than the radio supply.

The power dissipation in a switch is low, but the current is quite high, and

Component identification and handling

typical small plastic switching transistors up to about 1 A need no heatsinking. In the case of one used to handle the entire radio current when remotely switched on, a TO–220 PNP device capable of 10 A may be used.

8 Modulator AF stages in AM transmitters

Modulator AF stages in AM transmitters are most usually the same power AF stages which handle the AF to the loudspeaker during receive. A very few AM transmitter designs use a modulator device in series with the RF PA stage which does away with the modulation transformer but has disadvantages; for example it has to operate in Class A and hence have half the supply across it during zero modulation. This restricts available RF power, and the efficiency of the system is much lower than one using a transformer. Modulator devices are thus normally used in push–pull Class B with some quiescent current to reduce cross-over distortion and associated with a transformer. Plastic TO–220 packages are usual, bolted to the radio metal frame or to the PA heatsink. They can be NPN or PNP.

9 Simple low signal amplification

Simple low signal amplification may be performed with a discrete device where there are no spare OP–amps in a multiple chip, or for low noise AF uses such as a microphone pre-amplifier. JFETs, PNP and NPN devices are all used in this application and are all small and plastic. Other small signal applications include detectors, where an NPN device is held 'on' by a large resistor from the supply line to base, relying on incoming AF or IF signal to switch it 'off'. Thus the average collector DC voltage will climb as a function of input AC signal level. A typical circuit is given in Figure 10.10.

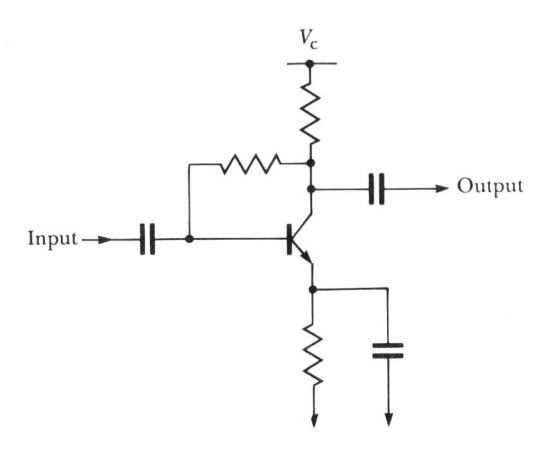

Figure 10.10

Such circuits are used to detect noise in squelch circuits, and to provide a

Mobile Radio Servicing Handbook

signal strength output from energy in the IF stages of a receiver. They are not particularly accurate or stable with temperature, typically their output varies by ±2 dB over −10 to +55 degrees centigrade for a constant AF input.

10 *Diodes*

Diodes are used in a variety of applications, some are listed below:

a Reverse current protection; this is a series diode in the supply line to the radio and has to handle the entire radio current, say 8 A. Its volt drop can be problematical as discussed in 7 above, in association with regulators, as the combined junction voltage and bulk semiconductor material resistance can result in a 1 V drop. A schottky diode can be used here to advantage, with its 0.1 V junction drop. Also, the maximum reverse voltage rating (VRRM) of the diode need only be, say, 25 V. The package will be bulky, either stud mount or a large plastic or ceramic body with thick leads, in order to achieve the required current rating. As its dissipation may be a few watts, it may well be mounted on a heatsink.

b Reverse voltage protection; this is a shunt diode used in association with a fuse. When the voltage is applied backwards, the diode conducts and the fuse blows. This has the advantage that there is no volt drop in the supply line, but until the fuse blows there exists some 1 V across the radio in the reverse direction, which is more than enough to damage ICs. However, this method remains the most popular protection against wrong supply polarity. The package is large, plastic or ceramic with quite thick leads, in order to present a low impedance to blow the fuse quickly. Its dissipation is zero under normal conditions, so it will not be on a heatsink.

c RF detection; most FM transmitters have a control loop around the PA to keep the RF output power constant over a range of supply voltages and ambient temperatures. A shottky diode is often used to detect a 'sniff' of the RF output as part of the loop. This is a small glass encapsulated device.

d AF detection; the humbler germanium glass encapsulated diode is still popular for AF detection in AM receivers. Active AF detectors may also be found in ALC or compressor stages in transmitter audio processor circuits, or in noise detectors in FM receiver squelch circuits, using silicon glass encapsulated diodes.

e Limiters; silicon diodes in the feedback around OP-amps are commonly used as AF limiter circuits, to be found where, for example, an audio signal enters an IC such as a modem. Other applications for limiters are in ignition noise suppression where the amplitude of the limited signal is altered to the average value of the AF passing through it, or the limiting of an IF signal just prior to a second mixer in an FM receiver.

f RF Switching; PIN diodes are commonly used to form an RF switch to connect the receiver or the transmitter to the harmonic filter and hence the antenna. They are also used as variable resistance devices in phase modulators in crystalled transmitters. The packages are usually small and glass for modulator PIN diodes, and small and ceramic with thick legs for power RF switching. Power dissipation can be a couple of hundred milliwatts for a PIN diode in series with the transmitter RF.

Integrated circuits

Integrated circuits come in two basic types, analogue and digital. The varieties of each commonly found in radios are:

Analogue types

a OP-amps. Usually in twins or quads for audio and DC applications.
b IF amplifiers/limiters and FM detectors.
c AF output stages.

Digital types

a Transmit–receive switching logic and timing.
b Microprocessor devices.

Digital types are usually CMOS with their usual problems of being damaged by static discharges. Analogue types are generally more rugged, although some OP–amps have CMOS FET input devices. It is important to look up an IC in a data book if any doubt exists as to its function or to its handling.

ICs come in various packages. Some older radios using the first available analogue ICs may contain metal round ones with up to twelve legs. Most modern radios have ICs which are dual-in-line (DIL) or single-in-line (SIL), which makes PCB layout design and fitting easier. The majority of ICs used will be plastic; the ceramic ones have wider operating temperature range and are more expensive; they are able to operate to $-30°C$. More and more radios use a 'condensed' DIL package with tabs spaced at 1.5 mm and the IC mounted on the track side of the PCB.

Passive components

The passive components discussed here fall into three categories – inductors, capacitors and resistors.

Inductors come in two basic types:

a Chokes; these are found in series with supply lines to prevent the passage of AC signals above a certain frequency. They may also be used to hold a point at DC ground while permitting AC voltage above a certain frequency to

Mobile Radio Servicing Handbook

appear at that point. An example is the base of an RF PA transistor, where the rectification effect of the base-emitter junction must not be allowed to bias the device off. A choke is normally placed between base and ground. The AC frequency at which an inductor becomes active as a choke is below its resonant frequency when considering both its self capacitance and the circuit capacitance. The packages are either small wound bobbins, perhaps encapsulated in resin, or wound around ferrite beads or rings. Some encapsulated bobbins are colour coded much like resistors.

b Resonant inductors; whereas chokes are inductors used below resonance, some inductors are deliberately operated at or near resonance for frequency selection. These inductors may be fixed or variable, and can consist of a length of PC track with tight tolerance dimensions, but portrayed on the circuit diagram as an inductor. Interstage RF amplifier matching in transmitters is a favourite location for these, and some multi-stage receive filters have inductive coupling using PC track. It all depends on the value of inductance required. If very low, say to 50 nH, then using PC track is simplest and cheapest. Above this value, say to a few hundred nanohenries, air wound stiff wire coils on top of the PCB will suffice. Variable inductors will be wound on a former with a ferrite tuning slug within it, and probably a screening can. Values of inductance greater than this are not normally required on mobile radio, as the minimum frequency covered is of the order of tens of MHz, and IF filtering is normally done using crystal or ceramic devices.

The only failure mode with inductors is when a choke burns out due to excessive DC current. However, ferrite can and does crack, which can impede the performance of a choke, or render a variable inductor useless. When adjusting slugs in variable inductors, the most common cause of slug fracture is use of the wrong tool. Replacement of the whole assembly is then often necessary unless the slug can be drilled out without damaging the former.

Chip inductors do exist, but they are usually a wound chip bobbin and quite expensive and rare on mobile radios. Their inherent low Q makes them useless for selective filters, but they may find application as chokes in decoupling circuits.

Capacitors come in many types. Listed below are some popular ones used in mobile radios and their application.

a Disc ceramic – values of 1 nF to 100 nF. General decoupling capacitors. Their tolerance and unavoidable minimum lead length makes them unsuitable for filters or matching circuits.

b Plate ceramic – values of 0.5 pF to 1 nF. Used in filters up to Band 3, and some RF interstage matching. Also, RF decoupling for UHF. Their tolerance is quite good (10 per cent) and they have less minimum lead length than disc ceramics.

c Cased mica – values of 1 pF to 1 nF. Used in filter and RF matching circuits up to and including UHF. They are square in construction, either 5 mm or 10 mm square with a tab to connect to one side of the capacitor and the body to connect to the other. These capacitors have virtually no associated inductance and their loss angle and tolerance (5 per cent) are very good. They are moderately expensive and take up a considerable amount of room on the PCB. They need to be fitted by hand. However, there are areas of the radio where no other type will suffice, such as the PA harmonic filter.

d Mylar – these are close tolerance (5 per cent) capacitors with a value range of 1 nF to 220 nF or so, and are used in AF active and passive filters whenever a comparatively large value, close tolerance capacitor is required.

e Electrolytics – these are used whenever the capacitance value required exceeds about 470 nF. They come in various types. For values up to about 22 μF capacitors with tantalum electrodes are often used because they are smaller, but above this value they become uneconomical and aluminium types can take over. The maximum value likely to be encountered is 1000 μF. Aluminium electrolytics have a huge tolerance on their value since the value you get depends upon the voltage across the capacitor. Typically it is -80 to $+100$ per cent. The value is correct only when the capacitor is working at its rated voltage, which means that there is never much headroom, 16 V electrolytics are found on sets working from 13.8 V, so, the electrolytics become casualties when the supply voltage is taken outside specification.

f Variable capacitors – all variable types will be meshing vanes with a dielectric of PTFE or of air, according to the value and Q required. These are used in filters and matching circuits usually in conjunction with a small fixed inductor.

Of all these types, ceramic capacitors are the ones available in chip form, although these can be multi-layer to give a total value range of some 0.5 pF to 100 nF or more. If values less than 0.5 pF are required, then a capacitor is normally formed using copper areas of the PCB adjacent to one another.

Capacitors rarely fail. They may short-circuit if the dielectric is overstressed with excessive RF or DC voltages, in which case the failure is catastrophic and only too obvious. Cased mica and chip capacitors usually show scorch marks in this condition. Chip capacitors sometimes fail mechanically if the PCB is stressed. The end-caps can part company with the body or the body can split with a hair-line fracture. A diode-probe RF voltmeter is the best instrument for locating such discontinuities.

Resistors are possibly the most common discrete components to be found in electronic equipments. Their function is to resist or impede the current flow in a circuit. The value of the resistor is defined in ohms and its value is either written on the body or more commonly denoted by a colour coded marking system. The resistor will have an accuracy tolerance to its nominal value and this will be likewise marked. As it resists current flow, heat will be generated and so a resistor needs to be of sufficient wattage rating for the task. An approximation to its wattage may be assumed from its size although the resistor construction and material will also affect this. Resistors may be wirewound (not usually suitable for RF due to their inductance), carbon composition, metal film, high stability carbon film and thick film. The value of a resistor is read by starting at the end opposite to the broader band and noting the colour bands in order. For example:

- Band 1 – brown
- Band 2 – red
- Band 3 – yellow
- Band 4 – (broad band) red

Band 1 is noted from the colour table e.g. 1. Band 2 is noted from the colour table e.g. 2. Band 3 is noted from the colour table but instead of being written directly is used to indicate how many zeros follow the first two bands, i.e. yellow = 4, therefore 4 zeros e.g. 120000 ohms or, 120 K ohms. Band 4 denotes the tolerance, e.g. 2 per cent. Note that black means no zeros.

Note: This is an example of one type of resistor coding only, but other systems use more or less bands and different colours for tolerance etc.

Colour Code Table

Black	0
Brown	1
Red	2
Orange	3
Yellow	4
Green	5
Blue	6
Violet	7
Grey	8
White	9

Handling Semiconductors

Storage

Devices should be kept in a dry, room temperature environment, preferably in compartmented drawers, one compartment per device type. The only devices sensitive to static discharges are MOS. These should be kept in carbon-loaded plastic containers available for the purpose. Chip transistors and diodes come on self-adhesive tape, and have a number on them which refers to a look-up table from a particular manufacturer since it is impossible to fit the complete type number on the device. Integrated circuits should be stored in the packaging in which they arrive. The ideal storage medium is conductive plastic extrusion which is made for the purpose.

Handling

It is good practice to always handle devices with tweezers. In a general stores where devices are kept in bulk quantity, the store should issue devices in small numbers in individually labelled conductive plastic bags to prevent static damage. ICs should be issued in short sections of conductive plastic extrusion especially made for the purpose. Whereas most diodes and transistors are not affected by static, a growing number of semiconductor devices are so affected, and a good rule is to avoid static altogether, particularly when these points are not understood by non-technical stores personnel. When transferring devices from one place to another, try to have bodily contact with the original environment and the destination environment, and if transferring between people, touch that person when handing over the device. Remember that it is voltage across the device which does the damage, so walking across a nylon carpet in rubber shoes whilst carrying the device does no harm in itself, but then if the device is given to another person, movement of charge will take place and damage will result.

Removal from and replacement into a PCB

Once a device is removed from a PCB it is useless. This statement is bound to raise controversy, but in every workshop where second-hand components are kept they are not trusted, and certainly should not be used to repair another radio. So, the main concern in component removal is preservation of the PCB itself. Devices are mounted in one of two ways. They have legs which poke right through the PCB soldered top, bottom, or both, or they mount on one side of the PCB with tabs. The most difficult semiconductors to remove are PA transistors with tabs, and a stud which goes through the PCB and heatsink, secured with a nut. Most PCB designers make sure that the copper is brought right up to the body of the device.

With any device removal the first thing to do is to remove as much solder as possible. The performance of most solder-suckers is enhanced by fitting an expandable 3 mm or 4 mm sleeve over the tip. This is particularly recommended for plated-through hole PCBs. The flexible sleeve forms a seal around the hole, and the sucking action then cleans the hole right out. For devices with tabs, the tab with solder removed is then heated and curled away from the copper up to the device body. Remember that the tab will be soldered underneath.

When removing PA devices with studs, it is recommended that the tabs be unsoldered and peeled back before the stud nut is removed, as it is easy to push down on the stud while the tabs are still soldered and hence rip the copper track. Note that when copper on a PCB is heated, the adhesive which attaches it to the PCB material loosens, making the copper very easy to peel off.

With chip devices, the procedure is basically the same, except that there is generally no compliance in the device construction, and both ends of the device need heating and releasing from the PCB together after the excess solder has been removed. Chip ICs are more of a problem; a special large iron is required to release all the legs together, otherwise the body can be snipped off and the legs removed one by one.

Replacement is an easier process. Firstly clean up the PCB copper track and any holes with solder wick. If the device has tabs, especially large ones, tin the tabs with solder on their undersides where they will contact the PCB solder areas. Introduce the device to its correct position, and bolt it down if it has a stud. With leaded devices, solder around all the legs quite quickly, all within ten seconds if possible. Modern semiconductor devices do not need heat shunt as long as the soldering process is not long. On a device with tabs, heat the tabs such that the solder trapped between the tab and the PCB melts, and introduce more solder around the tab edge. When fitting chip components, hold the chip with tweezers while soldering one end, and then solder the other end. Chip components tend to let go of their end caps when heated too much, so again the rule is to solder as quickly as possible. Use as little solder as possible consistent with obtaining a good joint. This book is not about good soldering technique, but a few tips are given below:

a Use a low voltage temperature controlled iron. This type will not damage MOS devices and minimizes the risk of dry joints.
b Use 22 swg 60/40 solder. Thicker solder will result in too much being applied.
c Keep the iron bit clean with frequent wiping on a damp sponge.

Components mounted on top of the PCB with their legs going through it, should be mounted as close as possible to the PCB. This has particular relevance in RF or IF circuits where excessive lead length can lead to microphony or instability. The hole spacing on some PCBs tends to spread the device legs, and so damage may occur to the device if it is forced down onto the PCB. On the other hand, it is never a good idea to mount leaded transistors touching the PCB even if the hole spacing will allow it. A gap of 1 mm to 2 mm should be aimed at.

PCB track repair

Occasionally due to excess current or heat applied during repair, a PCB track becomes damaged. The first step in repair is to make a sketch of the route that the damaged track followed. Then cut away loose or hanging track back to the point where the track is still bonded firmly to the PCB. Carefully tin 4 or 5 mm of the remaining track using a fine bit and minimum heat so that more track does not lift. Using tweezers hold a piece of tinned copper wire (24–28 swg) and solder it to one end of the track. Without transmitting stress to the soldered track, form the wire along the route the track took previously and solder it at the other end. If the repaired track is more than approximately 20 mm long, it should be secured to the PCB with a very small blob of glue.

When working with PCBs having track faults, either tracks bridged with whiskers or tracks open circuit due to hairline cracks, it is useful to use a milliohm meter to localize the problem. Tracks that are electrically continuous, but lifted from the PCB may either be repaired as broken tracks or simply restuck to the PCB.

Flexible PCBs with damaged tracks may be repaired by using silver loaded paint.

Surface mount technology

Surface mounted technology (SMT) – the realization of electronic components mounted on the surface of a printed circuit board (PCB) rather than through it – produces benefits to both manufacturers and users.

a Increase in component density. Being smaller than their leaded counterparts, chip components allow much higher packing densities. It is quite possible to achieve a 50 per cent PCB area reduction when re-engineering with SMDs.
b SMDs can be used with conventional printed circuit board materials and soldering methods. This allows a mixing of leaded and surface mount technologies on the same PCB.
c Increased production automation. Many types of components (resistors, capacitors, inductors, etc.) were previously encapsulated in unique and differing package styles. In SMT these all take the same outline or 'footprint' and so can be assembled onto a PCB with just one standard machine.
d Increased reliability due to the reduced number of component interconnections.
e Both sides of the PCB can be used with equal ease.
f Decreased storage area requirements at manufacturing stages.
g Improved circuit performance.

While most of these benefits are enjoyed by large sectors of the electronics industry, it is the last point that needs to be considered in detail for the RF industry.

SMDs have given the RF design engineer much more versatility on conventional PCBs using common fibre glass material. As component interconnections are on one surface of the PCB, the RF characteristics of this interconnection or 'line' approaches closely to the ideal and are not spoilt by plated through holes etc. Using a continuous ground plane on the reverse side of the board, a required impedance line (often 50 ohms) with good terminations, can be implemented. The advantage of this is that the circuit will have a better, or more closely designed, frequency response and less unwanted radiation from the line due to mismatch. This will have the further advantage that it is accurately reproducible from batch to batch, as PCB manufacturers work to very tightly controlled tolerances; this last being necessary due to the advent of CAD/CAM techniques.

Reduced track lengths also mean less inductance and this reinforces the previous points.

Calibrating and servicing SMT equipment

Although we have outlined the advantages of SMT, it does demand new requirements and skills of the engineer, who is required to handle this technology.

The benefits discussed are obtained by utilising all of the modern design aids available. Computer techniques, both in PCB design/layout and circuit modelling, have meant that previous 'over engineered' safety margins no longer exist. The service engineer needs to work with a much greater awareness of the restrictions and rules placed on him. The job of the service engineer is to test the equipment and hence determine if there is a fault, to locate the fault and to rectify the fault, all without causing further problems or to degrade the original performance of the equipment. In order to achieve this, a degree of discipline is necessary in the procedures used and actions taken.

As stray capacitances are considerably reduced in SMD modules, the addition of test probe capacitances when fault finding have a proportionally much larger effect. This disturbance of the circuit may be sufficient to move the operation of the unit outside the available adjustment range of present alignment controls, or may even stop the circuit working completely. The approach to this problem is to first check the manufacturer's service manual for a recommended measurement test point. In the absence of such a test point the service engineer should examine the function of the circuit to determine whether there is a circuit mode that will be unaffected by the addition of probe loading effects. If there is doubt that such a point exists, then measurements should be made after the next circuit in the signal path in order to isolate the disturbance effects. Some assumptions will have to be made using this

Component identification and handling

technique, but in working along a chain any incorrect circuit actions will soon become apparent.

When probing around SMDs, particular care must be taken not to fracture the relatively delicate structure of these devices. Whereas with previous component technology some movement could be accommodated by lead 'flexibility', only fracture will allow movement with SMDs.

Certain precautions must be taken when working with, or replacing, SMDs.

a) The test probe tips should be applied only to the solder meniscus and not to the coated top of the device. If the device has not been mounted perfectly flat on the PCB, a force will be applied which will fracture the SMD.

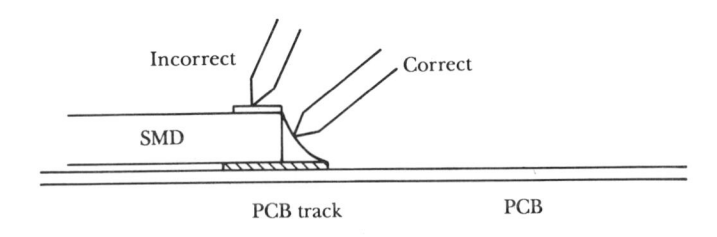

Figure 10.11

b) The manufacturer will certainly use an adhesive to stick the components to the PCB prior to soldering. If a device is to be changed, not only must the end terminations be unsoldered (usually by pump or wick), but the body must be prised off the PCB – usually in pieces. It is not necessary to use adhesive for rework – the device can be held in place with plastic tweezers or a fine plastic point. Care must be taken not to damage the environmental, coating or end terminations of the replacement device.

c) The replacement device is all important. It is a great temptation to believe that because the components are extremely small and often look identical, that the difference between one and another is not great. Isn't a ceramic capacitor just a ceramic capacitor? The comments with regard to computer design must be kept in mind – there is little or no room for component substitution. The ceramic capacitor may have any one of the usual dielectric materials, NPO, COG, X7R, Y5V etc, and circuit performance may depend for its stability on the correct material being used. Reference must be made to the manufacturer's parts list for a component description or type number. If in doubt the tightest tolerance/temperature stability material should be used.

d) There are different body sizes of SMDs. The pads on the PCB will have been optimised for a particular size and also stray or coupling capacity may vary with body size. Popular sizes are shown in Table 10.1.

Mobile Radio Servicing Handbook

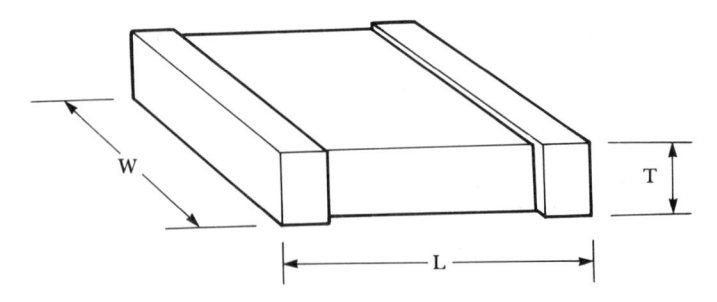

Figure 10.12

Table 10.1

Body Style	0805	1206	1210	1812
L	2.0	3.2	3.2	4.5
W	1.25	1.6	2.5	3.2
T max	1.27	1.65	1.91	1.91

If possible, replacement devices should be of the same body size as the original component.

e) An extremely important consideration for replacement devices is the termination material. The parameter of concern to the service engineer is that of LEACH RESISTANCE. The termination material of an SMD needs to adhere strongly to the ends of the device, and, to achieve this an alloy of silver and palladium (PdAg) is used. The problem is that molten solder, particularly if rich in silver, causes the dissolution of silver from the termination. This can, in turn, destroy the metallization and hence its bonding to the device being soldered. This undesirable process is known as leaching, and is both time and temperature dependent. Although this may not be a problem during manufacture, as both solder temperature and time are carefully controlled, in the service environment this will probably not be the case. Most SMDs are available with a nickel plated 'overcoat' on the termination and it is these that should be used as replacement devices. It is also important to use a good quality temperature controlled iron for re-work, a temperature of 260°C is usually recommended. The following graphs (Fig. 10.13 and 10.14) show the improvement available using the nickel plated terminations over naked palladium silver.

f) When tracing a circuit function on a PCB against the manufacturer's circuit diagram, certain components may appear to be missing. However, turning the board over will reveal them! Manufacturers make full use of the fact that

Component identification and handling

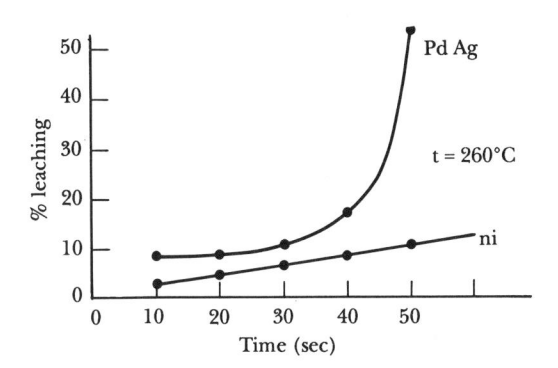

Figure 10.13

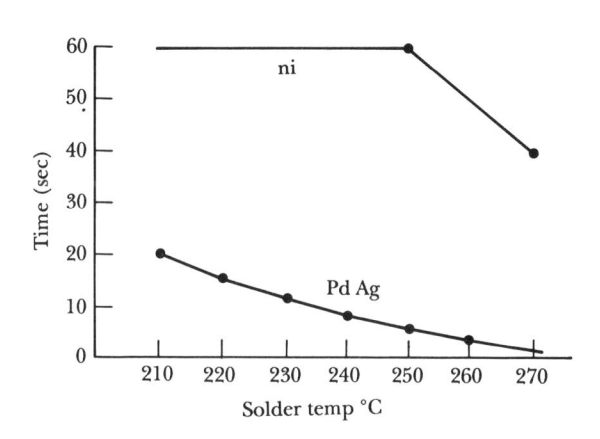

Figure 10.14

SMDs may be as easily mounted on one side of the PCB as the other. An obvious point, but one often overlooked.

Workshop practice

SMDs being small have either none or a minimum amount of information marked on them. Resistors are sometimes marked with a value but not with other information. The same comments apply to capacitors. Most service workshops will have a multimeter so an unidentified chip may be recognised as a resistor and its value measured. However, in the case of a capacitor this may well be a ceramic type with just a few picofarads capacitance. Capacitance meters measuring down to a few pf are expensive, need careful driving if stray capacitance is not to be a problem, and are not so often found in the small service workshop. All this means that prevention is better than cure. It is better to store safely and label clearly all SMDs. Even when purchasing small (50 pieces) quantities they are usually packaged in bandolier or strip form, which is convenient. These strips may be stored in labelled plastic-drawers or even polythene bags (watch the static problem) and as an added precaution it is worth marking the back of the strip with a marker pen with the value of the device. Finally, given the low unit cost of these devices and the cost of service engineering time tracking down a wrong value component, it is better to follow the maxim – 'If in doubt, throw it out'.

11 Electromagnetic compatibility (EMC)

Introduction

EMC has become much more of a problem recently with mobile radio, as microprocessors are now cheap enough to install in every one. The computer manufacturer will argue that all computers generate noise and it is up to the radio designer to screen the sensitive RF and IF stages of the receiver. The radio designer disagrees: all radios are sensitive to radiated noise so the computer designer must screen the computer. This argument will continue as long as companies use different departments to design the different areas of radio. Co-operation will succeed.

More electromagnetic interference may be caused by CMOS digital ICs, with their larger scale of integration.

Two aspects are considered here with respect to microprocessors within radios, reducing radiation and reducing sensitivity to radiation, and the last aspect considered is installation techniques in vehicles and base station sites to minimize potential interference.

Methods of reducing radiation

The layout of a microprocessor PCB is nearly always designed by computer aided design. A computer, when routing tracks carrying the data and address buses does not consider radiation problems unless some guidelines are given to it by the operator. These guidelines are:

1. Use tracks which are short and as fat as possible to achieve minimum inductance. The clock and bus lines carry signals into the MHz region, so these in particular should be kept short. The position and orientation of the LSI devices should optimize this. Most microprocessor PCBs tend to use long spidery tracks much less than 1 mm wide, which act as wonderful antennas.
2. Use as much ground plane as possible, top and bottom, with an abundance of track pins or PTH to join the two together. Multi-layer PCBs are now available with a ground plane sandwiched in the middle, but these are rather expensive. Various sprays are available which claim to put a screening layer

Mobile Radio Servicing Handbook

over the top of the components. Such devices should not be used in preference to getting the layout optimum at the beginning.

3 Allow provision on the layout for connection of the microprocessor ground to the radio ground, with as low a resistance as possible, preferably have both bolted to a common metal frame so that a minimum of potential difference exists between them. The power supply wiring is regarded as ground to AC, so frequent and thorough decoupling is essential.

Mount the microprocessor PCB away from RF and IF receiver stages, if possible. LSI digital chips radiate from their bodies, so there is some mileage in having the track side of the micro PCB towards the radio, in that a socketed LSI chip stands proud by some 8 mm.

The ideal way to package a microprocessor is in its own die-cast compartment within the radio with feed through capacitors on every lead. Figure 11.1 gives the general idea.

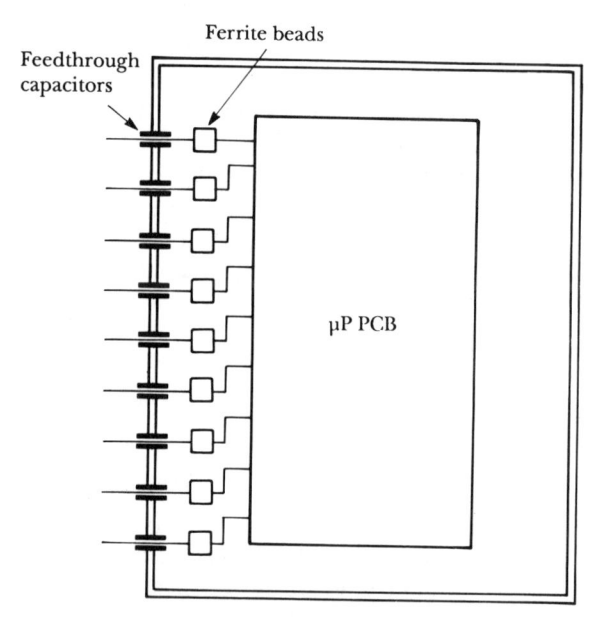

Figure 11.1 *Microprocessor total screening*

Feed throughs used in this way reduce radiation from leads by some 70 dB and by totally enclosing the PCB, a shorted turn is formed around it which causes the radiated energy to be lost in circulating currents. Simple electrostatic screens around part of the circuitry do not give effective protection.

Some radios have the microprocessor as part of the front assembly, with all of the sensitive RF and IF circuits at the back. This is better from the radiated interference viewpoint, but power has to be brought the length of the radio for the microprocessor, so adequate decoupling is essential.

What interference there is, is bound to be worse on some radio channels than on others, and may get worse or better, according to what the microprocessor is doing. It is dangerous to put the microprocessor in 'test mode' to check the radio and then check the interference levels. The microprocessor will be very quiet anyway in such a mode. If all attempts at ridding interference fail on certain channels, say with a spacing of 1 MHz, it is possible as a last ditch effort to move the microprocessor clock frequency slightly when on those channels to move the interfering spectral line a couple of channels away. The microprocessor could automatically switch a transistor which connects a capacitor across the clock crystal. This has complications, not least of which concerns the accuracy of the clock. It may be necessary to choose a clock crystal which is very accurate so that the interference appears on the same RF channels between radios so that the clock pulling always works. Also, suddenly changing the clock frequency may introduce a transient which upsets the microprocessor. The software writer should provide a flag to do this at a convenient time.

Methods of reducing sensitivity to radiation

It is always regarded as good practice to suppress interference at its source, but simple screening of a microprocessor is not possible unless it is totally enclosed and decoupled. There may, however, be areas of the radio which can be screened to advantage. These areas are:

1. The path from the antenna socket to the input of the RF amplifier in the receiver. This will go through the harmonic filter, some RF transmit/receive switching, and a bandpass input filter. Given that international RF radiation limits are so low (0.25 μW in Europe), it is likely that these areas will be screened already.
2. The mixer stage and path to the first IF filter and amplifier.
3. The VCO in the synthesizer. Noise appearing on the local oscillator will be translated into audio interference by the FM detector.

Measurements performed by the author indicate that 8-bit microprocessors running with a 4 MHz clock, generate noise stretching from HF right up to 300 MHz or so, encompassing VHF bands and Band 3. The contributions to the interference at the output of the receiver from each of the RF stage, mixer stage, and that picked up in the IF, are approximately equal. Therefore the screening of all the RF and IF circuitry is desirable.

Mobile Radio Servicing Handbook
Installation techniques to minimize interference

Mobiles in vehicles

The ground rules are:

1 Use double screened coax from the radio to the antenna. This minimizes the potential difference between the car metalwork and the radio chassis. Use an antenna base which grips to the metalwork of the car, not a magmount. Try and mount the antenna in the centre of the roof. This keeps it away from engines on other cars and bikes.
2 Mount the radio as far as possible from the ignition coil and distributor, and route all leads from the radio away from these. Keep the supply leads running with the antenna coax for as long as possible, to keep the enclosed area between them down. Figure 11.2 illustrates what happens when this enclosed area is large, particularly when there is an interfering source within the loop.

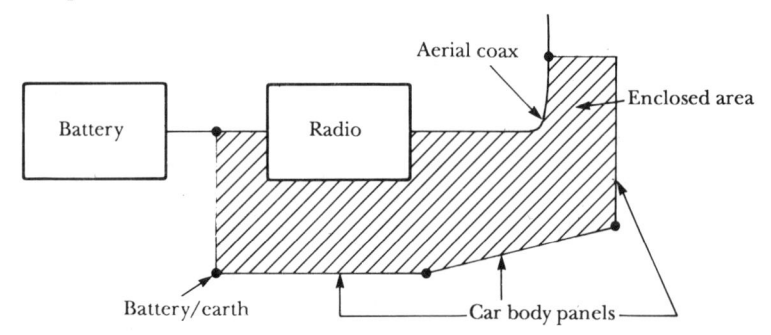

Figure 11.2 *Generation of closed loop on the ground of an installation*

3 Ground the bonnet lid. Although this may be inconvenient and seemingly irrelevant, it does improve matters.

Various people hold that grounding the radio chassis to the car chassis directly improves rejection, but this is difficult as dashboards are now plastic and the author has tried it without much improvement.

There is a document (MPT specification available from the DTI Radio Division library, Waterloo Bridge House) a code of practice, published by the Department of Trade and Industry, which lays down guidelines for installation procedures and this should be followed.

Base stations

Base station installations do not usually have ignition interference to cope with, but on a site where several base stations coexist, unfortunate frequency mixes

Electromagnetic compatibility (EMC)

often occur. Antennas are dealt with in their own chapter; under discussion here are the mechanisms by which interfering signals can be generated and their elimination. The two areas where two signals or more can mix to produce other signals (intermodulation) are:

1. In the transmitter PA. Signals entering the PA from the antenna system will mix with the transmitted signal from the PA to produce sum and difference signals only about 15 dB down with respect to the interfering signal level. The PA device is running in Class C which makes it highly non-linear.
2. In the receiver front end. Intermodulation rejection is a critical aspect of receiver front end design. Modern base station receivers achieve a rejection ratio of 80 dB or better, but if the interfering signal is a couple of MHz away, additional bandpass or even notch filtering in the receive path may alleviate the problem.

On any site with more than one transmitter per band, it is recommended that isolators be fitted in series with the RF output from each transmitter. These ferrite devices will divert any signals coming back at the transmitter from the antenna system into a matched dummy load. On most busy sites, a single

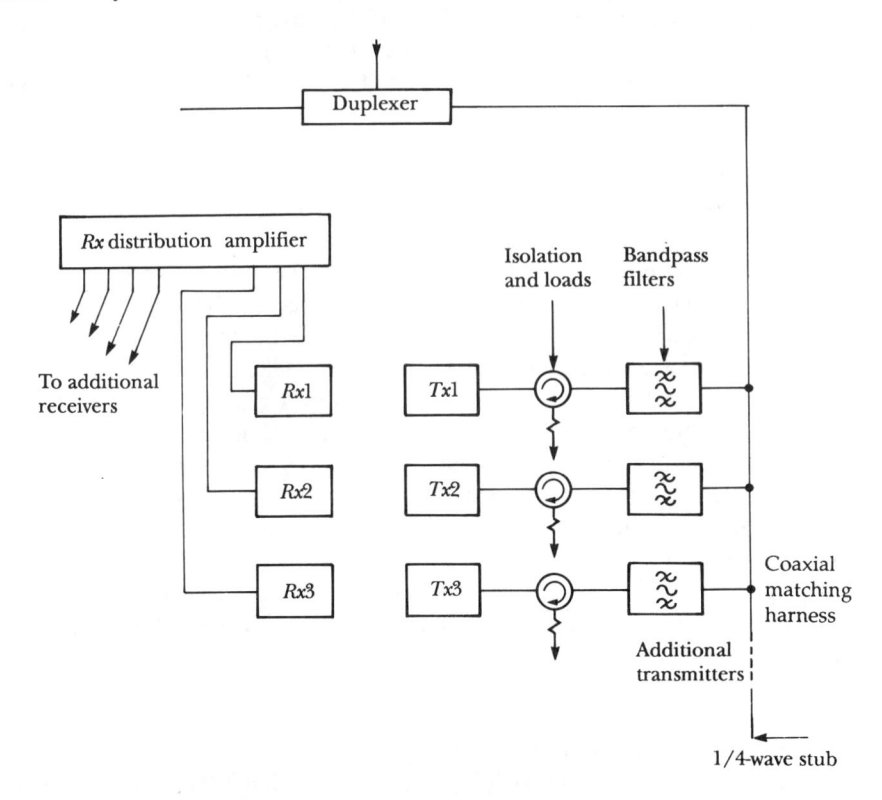

Figure 11.3 *Multicoupling several receivers and transmitters*

antenna is multi-coupled to each transmitter and receiver on a particular band. Each transmitter will have a bandpass filter in its path as well as an isolator; the band width of the filters will determine how close together the transmitter frequencies can be without interference problems. The aim is 20 dB isolation between transmitters using filters which are coupled to one another on their output sides and to a duplexer with a matching harness. From the other duplexer connection, a receiver distribution amplifier is fitted. Figure 11.3 gives a typical arrangement for five base stations multi-coupled to a single antenna.

Base stations are controlled by various means. A line operated base station will need connection to a 600 ohms two or four wire circuit. Most are two wire, i.e. both transmit and receive audio pass along a single pair of conductors. Additional signals are passed down the pair to control the base station; three control functions are generally needed:

1 transmit key;
2 talkthrough off;
3 CTCSS defeat (sometimes called tonelock).

The signal for talkthrough off will obviously not get through if the line is disconnected for any reason and the base station will go to talk through. This fail-safe condition is known as 'line fail talkthrough'. Recent legislation demands that the transmitter key be under the control of both the receiver squelch output, and a positive decode of a coherent signal such as CTCSS. Thus, both a valid CTCSS tone must be received and the receiver squelch must have lifted before the base station will transmit in talkthrough. The signal appearing at the operator's console must also be subject to a positive CTCSS decode, but there must be means to temporarily defeat this so that the operator can monitor the channel before he transmits, hence condition 3 above. It is usual for the received audio to be passed down the line at any time there is a valid CTCSS tone decoded, even when switched to talkthrough. However, the audio signal from the receiver must be prevented from reaching the transmitter when the remote operator is transmitting down the line. This is so that the signal from the operator console is not corrupted by signals received into the base station. Therefore the transmit key instruction from the console must override the talkthrough instruction. With 4 wire systems, separate pairs are used for receive and transmit audio signals which makes audio switching more difficult. AC or DC signalling systems may be used in theory, in practice AC signalling is most common where a British Telecom line is involved as BT do not guarantee a DC path. DC signalling equipment is cheaper and may be of use on very short private lines such as within an office block.

A community base station (CBS) is where several users share a common RF channel and each user group has a CTCSS tone allocated to it. Users not involved have their transmitters inhibited. This type of base station is the most popular at the moment. The various conditions a mobile can be in are as follows:

Electromagnetic compatibility (EMC)

Base station condition	Able to transmit	Mobile Condition
Free	Yes	No busy light
Allocated to this group	Yes	Busy light on and receiving audio
Allocated to another group	No	Busy light on but audio inhibited

The potential for interference is no higher than that for a base station dedicated to a single user, since only one user group is enabled at a time. Two mobiles can transmit together, of course, but the base receiver IF will limit on the strongest one in an effect known as capture.

In a trunked base station, the incoming call requests are queued and the co-sited base station channels allocated as they become free, much like a single queue system in a bank or post office. The channel spacing between say five trunked base transmitters is the same as that between their associated receivers which leads to intermodulation problems, as discussed earlier. Tracing interference problems can be more difficult since no user group is consistently allocated to one particular RF channel. A trunked base station is liable to have a computer associated with it, which brings us back to the EMC problems between radios and computers discussed earlier and the same basic ground rules apply.

Sources of radio interference

Radio interference can be generated by:

Electrical Interference	Car ignition
	Electric motors
	Street lighting
	Industrial, scientific or medical equipment
Atmospheric	Long distance reception (freak conditions) causing co-channel interference
Intermodulation	The mixing of two or more frequencies, power amplifiers of base station transmitters are a common source, rusty bolts in antenna systems can create similar problems.

When a user has an interference problem, he should contact the local radio investigation service office (see Appendix).

It is important to differentiate between remedial and enforcement action. Remedial action is where problems are caused by other licensed users acting in accordance with their licences. If you call in the UK Radio Investigation

Service, their time is charged at a commercial rate. Enforcement is where problems are caused by unauthorized use of radio and/or frequency assignment. In these instances, only a diagnostic charge is levied by the RIS (one hour's time charged).

Typical action which could be taken by a service provider could include:

Reduction in antenna height to reduce co-channel interference
The use of intermodulation devices such as circulators.

It is important that equipment within a system is regularly maintained to avoid frequency drift or spurious signal generation. Users should also beware of second hand equipment which may not conform with current DTI requirements.

Whichever interference is experienced, it is well worth keeping a record of the interference type, time and duration, together with relevant call sign information.

Broadcasting transmitters

Very high power broadcasting transmitters (500 KW and above) are capable of inducing millivolts of signal into inadequately shielded cable in which land mobile transmission may be working at microvolts – a problem to be aware of on shared broadcast/PMR sites.

Strong fields, of 50–100 volts per metre are often present 30–40 metres down in a building with, say, a high power broadcast transmitter installed on the roof. These RF levels may well be sufficient to upset electronic machines such as cash registers, if RF shielding is insufficient.

EEC directive April 1988

In April 1988, the EEC issued a directive whereby EMC limits will become statutory on a Pan-European basis from January 1990 onwards.

12 Antennas – selection, installation, fault finding and maintenance

Introduction

The UK shares the international problem of a limited amount of available and suitable frequency spectrum required to support a fast expanding user base. The solution traditionally applied is to reduce channel spacing, increase channel loading and work towards increasing frequency re-use within relatively close geographic proximity. Planning constraints limit the availability of new sites for radio masts – additional coverage and denser usage patterns have to be supported by better use of existing facilities.

Choosing an antenna which provides optimum performance for a given coverage requirement, siting, correct installation and maintenance of the antenna through its operational life, have all become increasingly important. This chapter reviews the choice of antennas now generally available and offers guidance on selection, application, installation, fault diagnosis and maintenance support techniques.

Note, this chapter refers to antennas as opposed to aerials, for all practical purposes the terms may be regarded as interchangeable.

Antenna gain and patterns

An antenna is a passive device radiating the same amount of power supplied to it by the transmitter or supplying the receiver with the same amount of signal collected from the atmosphere.

Antennas take many forms, each with their own applications or uses. The most simple form an antenna can take is the half wave dipole, either centre or end fed. The radiation pattern this produces is the classic doughnut shape (Figure 12.1) when viewed in 3D. The antenna has maximum radiation at its centre at ninety degrees to its length and minimum radiation at the ends. Halving the doughnut in both planes creates a figure of eight pattern in one plane and an omnidirectional circle in the other. These are referred to as the 'E plane' and the 'H plane' respectively, being the electric and magnetic components of the whole.

Mobile Radio Servicing Handbook

The E and H plane patterns are used for all antennas to describe the radiation pattern it will produce. Three dimensional patterns are not generally shown in manufacturers' catalogues, being difficult to produce. It is also worth noting that the radiation patterns shown in catalogues, unless stated, are shown in the best possible conditions, i.e. with the antenna in free space. This means there is nothing around or near the test antenna to affect its performance and we will see later that this is very rarely the case in practice.

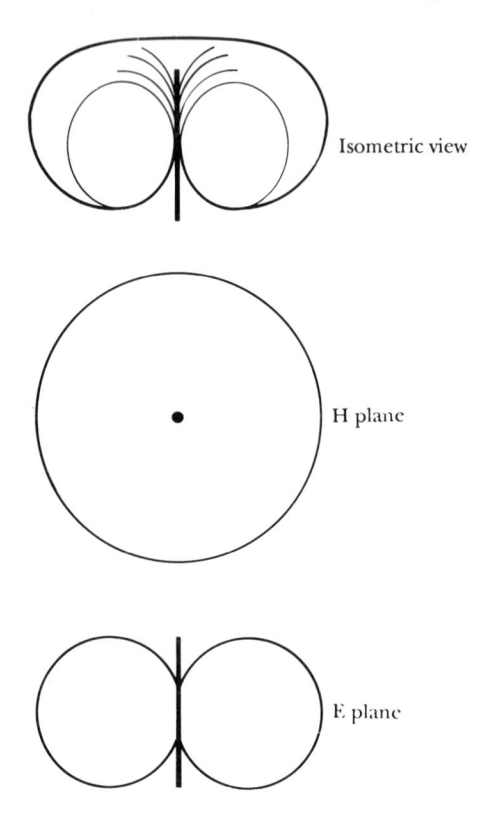

Isometric view

H plane

E plane

Figure 12.1 *Radiation patterns of a half wave dipole*

The direction of the electric field is known as the polarization, i.e. a vertical dipole produces a vertically polarized field. The receiving antenna must be similarly polarized. Normally transmissions are either vertically or horizontally polarized. Omni-polarization normally referred to as elliptical polarization may be produced by feeding vertically and horizontally polarized systems in quadrature, producing a rotating electric field.

The half wave dipole, being the simplest form an antenna can take, is used as a reference for gain or performance of other antennas. It has a unity gain or 0 dB. An antenna cannot increase the power applied to it. It is only a passive device. The only way we can achieve gain is to reshape the radiation pattern to put radiated signal where we want it. You cannot get something for nothing.

Directional antennas

The most common form of directional antenna is the yagi. In its most basic form it has two components, the driven element (dipole) and the reflector. The reflector is generally larger than the dipole and when placed the appropriate distance behind the dipole, pushes the signal normally radiated to the rear, towards the front, giving a gain improvement of 3 dB. This main lobe can be shaped further by adding elements or directors in front of the dipole. These are normally shorter than the dipole and with the appropriate spacing will narrow

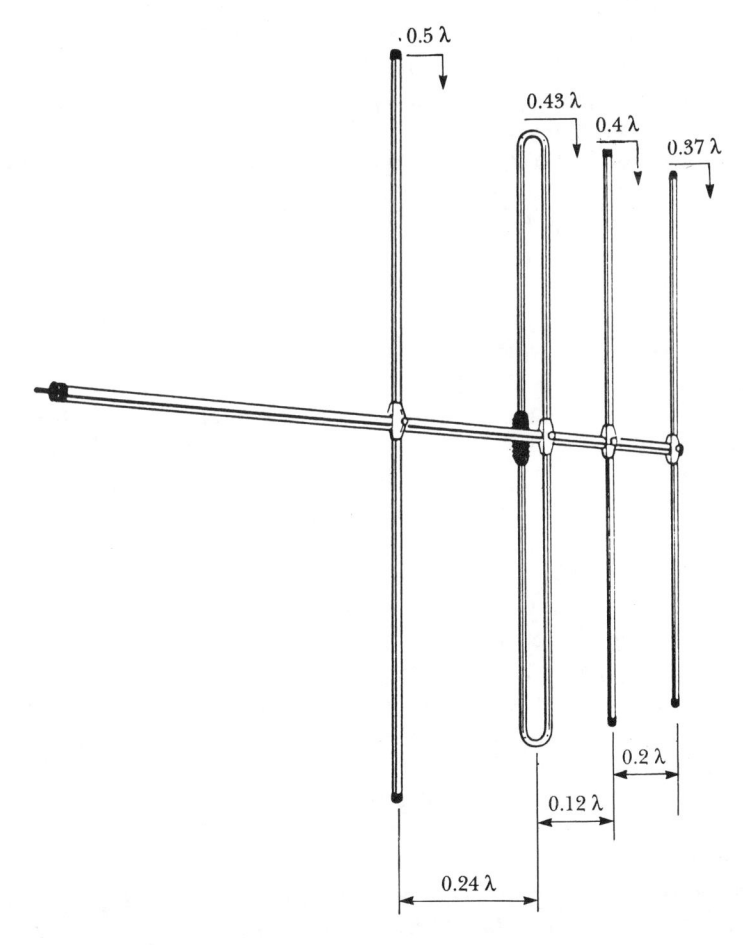

Figure 12.2 *Typical sizes and spacings of a four element yagi*

the main beam further, making the antenna more directional and thus giving higher gain. It can be said that the higher the gain, the narrower the beamwidth and to achieve a 3 dB gain improvement, it is a fair guide that the antenna length needs to be doubled.

While there are written specifications defining RF performance (the MPT standards for telemetry and microwave, for example), there are no official guidelines on the mechanical design and construction of antennas. In practice antenna design is dictated by the mathematical relationship between frequency and wavelength, and the fundamental propagation theory of electromagnetics. Most antenna designs in use today are based on the design principles evolved by Hertz and Kessler in the late nineteenth and early twentieth century.

Figure 12.2 shows a typical yagi antenna with sizes and spacings of its elements shown in wavelengths. These sizes are best found during practical development. They are chosen to give the best forward gain whilst making sure the spurious side lobes away from the main lobe are as small as possible, also that the other parameters such as front to back ratio and impedance are correct. To an extent, it is trial and error. However, once the optimum sizes are found for a particular antenna on a particular frequency, they can then be directly scaled up and down in frequency to produce other frequency versions of the same style of antenna. This generally holds true for most antennas with minor modifications once scaled.

The only constraint to the number of directors you can put on an antenna is the final mechanical length, overall weight and wind loading. Also the more elements an antenna has, the narrower its band width, making it less useful as a standard antenna for a typical frequency band.

The maximum practical sizes of directional yagis available are normally a four element for lowband frequencies, an eight element for highband frequencies, and an eighteen element for UHF frequencies with twenty-four elements being used on large UHF TV antennas for long distance reception, for example on North Sea oil rigs. Twenty-four elements are also used on SHF yagis, 1427–1530 MHz band is an example. Due to the small size and diameter of the elements at this frequency it is normal practice to encapsulate the entire antenna in a shroud transparent to radio waves, e.g. glass fibre or similar. This avoids the delicate antenna becoming easily damaged. The shroud also provides protection against static interference from rain.

Selecting the yagi most suited to the application is therefore a function of trading forward coverage against beam width against size constraint, whilst taking into account the frequency and band width required. Typical applications in which yagi style antennas are used are, for example, a VHF high band three element yagi used as a fixed base station repeater; a UHF twelve element used for point to point telemetry or directed down a railway track to provide an on-board train radio link; an SHF twenty-four element shrouded yagi (where short distances allow) used as a point to point link as an alternative to a parabolic dish, with the advantage of lower cost and lighter

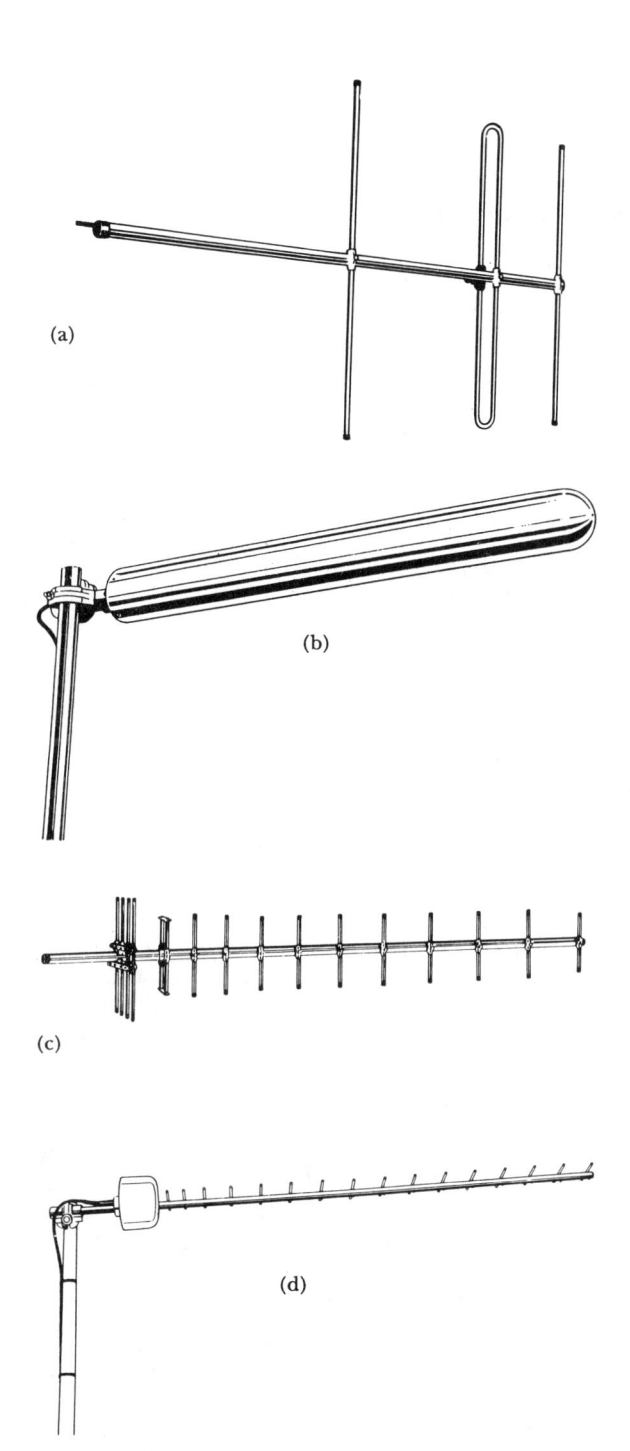

Figure 12.3 *Typical styles of yagi: (a) three element yagi at VHF frequency, (b) shrouded twenty-four element yagi at 1.5 GHz, (c) twelve element yagi at UHF frequency, (d) eighteen element yagi at centre frequency of 910 MHz*

loading on the supporting structures.

On the practical side, the larger the antenna chosen, the better supported it must be; the higher the gain, the narrower the beam width. Aiming/positioning accuracy at time of commissioning will be critical.

Table 12.1 Application guide – yagi directional antennas

Type	Peak gain (dBd)	beam width (degs)	E and H (degs)	Typical applications
2 element	3	62	95	Lowband frequencies
3 element	6	60	80	Highband VHF repeater link
4 element	7.5	58	74	Maximum size for lowband frequencies
6 element	8.5	56	64	High gain highband
8 element	10	40	50	Maximum size for VHF frequencies
12 element	12	30	36	UHF and above, point to point
24 element	15	25	26	SHF and above, point to point

Table 12.1 is a typical list of the types of yagis available, but as there are no guidelines on manufacturing antennas, the number of elements available and the performance will differ slightly from manufacturer to manufacturer and individual catalogues should be compared.

Stacked and bayed yagis

As has been seen, a point is reached where it is not practical to have a longer antenna. If additional gain is required, an alternative option is to place two identical antennas side by side (baying), or one over the other (stacking). The antennas must be identically orientated to each other and joined with a phasing harness. The phasing harness connects the two antennas into a single down lead, and ensures the correct impedance appears at the output, i.e. if two 50 ohm antennas are coupled together, 50 ohms is seen at the down lead junction. This is achieved in various ways, but most commonly by employing a quarter wave transforming section at the joint.

Doubling the length of the antenna gives an extra 3 dB of gain, doubling the number of antennas has the same effect. Every time the number of antennas is doubled 3 dB extra gain is achieved, so stacking and baying four antennas will give an extra 6 dB over the gain of a single antenna, stacking and baying eight antennas will give an extra 9 dBs. A point is reached where the loss in the phasing harness is the same or greater than the gain improvement, so it is not advisable to go beyond eight antennas without checking total cable loss.

To achieve this additional gain, the radiation pattern must be altered. If two antennas are stacked the vertical beam width is approximately halved. If two antennas are bayed, the horizontal beam width is approximately halved. It follows, if four antennas are stacked and bayed both beam widths are approximately halved.

Table 12.2 Stacking and baying yagis

Frequency band	Centre frequency (MHz)	Spacing (wavelengths)	Spacing (m)
Lowband	80	1	3.75
Midband	120	1.5	3.75
Highband	160	1.5	2.81
UHF 400–470	460	2.0	1.3
UHF 790–960	850	2.5	0.88
SHF	1500	2.5	0.50

It is important that the antennas are placed the correct distance apart to achieve the expected gain improvement. Table 12.2 shows the recommended spacing at various frequencies, and Figure 12.4 a typical arrangement.

It is also important to remember that, as before, the higher the gain, the less the usable band width of the antennas; the more antennas combined, the narrower the band covered.

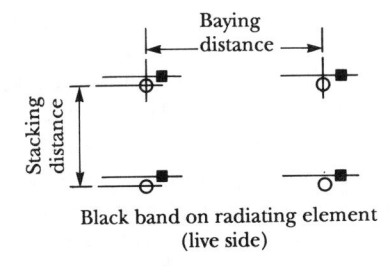

Figure 12.4 *Typical arrangement for stacking and baying directional antennas*

A side effect of phasing antennas is that a deep null is generated in the radiation pattern either side of the main forward beam. These nulls can be

Mobile Radio Servicing Handbook

swung around the radiation pattern independent of the main lobe. This is achieved by varying the spacing between the antennas. This can be useful for blocking an interfering signal by keeping the main beam in the direction required and altering the spacing of the antennas to point a null in the direction of the unwanted signal. This should not be confused with getting extra gain, as the spacing required to move the null might be too close to get the full gain improvement. Figure 12.5 shows ideal spacing for the null response, however adjustment in the field might be necessary for maximum rejection.

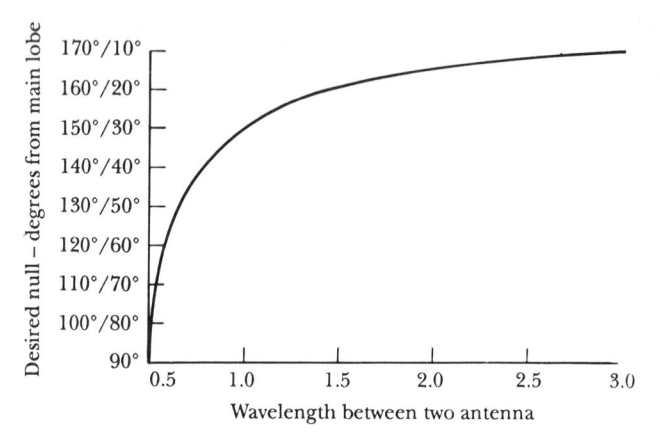

Figure 12.5 *Spacing of two antennas to achieve null response*

Other styles of directional antennas

Directional antennas for high front to back ratios

Front to back ratios, a measurement of how well a directional antenna discriminates between signals from the required direction and those directly behind it, are becoming increasingly important in those parts of the spectrum where users are in relatively close proximity, operating at similar or identical frequencies. A current example is the DTI requirement for a minimum front to back ratio of 20 dB for antennas used in the commercial Band 3 frequencies (175–225 MHz) in coastal areas adjacent to Southern Ireland, Belgium and France where Band 3 frequencies are still used for television transmissions and interference in both directions must be avoided.

The most effective way of improving front to back ratios is to replace the single reflector rod with a screen (Figure 12.6). This will give the same performance as a two element yagi but with a greatly improved rejection to the rear. Ideally the screen should be solid but this is impractical as the additional weight and wind loading is not desirable. The normal practice is to build the screen from rods with approximately 1.0 wave length spacing between them.

Antennas – selection, installation, fault finding and maintenance

At the higher frequencies of UHF this is quite easy to do with a range of yagi lengths, but at lower frequencies, the screen is more suited by baying two dipoles in front of it (giving 6 dBd gain), or by placing a slot style driven element in front of it, giving approximately 8.5 dBd gain.

The slot panel antenna can simply be described as three dipoles in front of a screen, fed top and bottom by a transmission line.

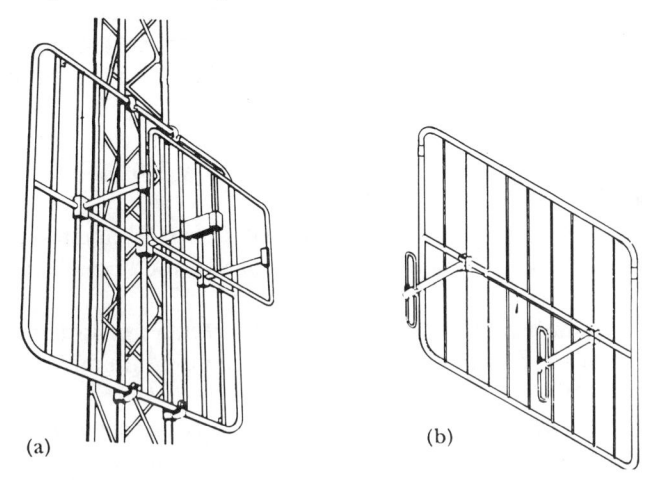

Figure 12.6 *(a) Slot panel antenna at VHF frequency, (b) Baying two dipoles in front of a screen*

To improve the gain of an antenna whilst maintaining a good front to back ratio, the flat screen with single dipole can be taken and the sides folded round

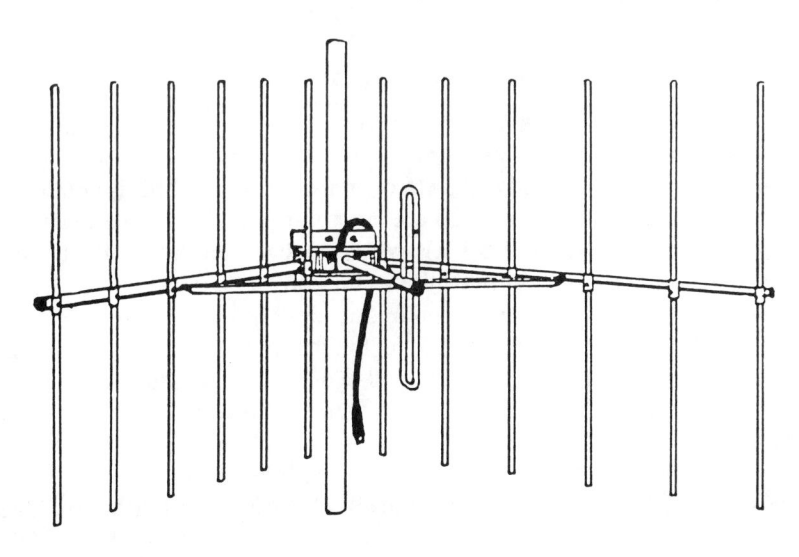

Figure 12.7 *Corner reflector antenna*

to form a vee, called the corner reflector antenna. This has the effect of narrowing the beam, increasing the gain. The resultant beam width and gain of the corner reflector depends on the size of the screen, the angle of the sides to each other, and the distance of the dipole from the screen.

Corner reflectors are normally designed to give a particular beam width as opposed to gain, for example a sixty degree beam width is common for sector coverage on main stations in telemetry schemes, or for covering a particular arc in a mobile radio scheme. If we take the screen one step further and form it into a parabolic curve in both directions we have the maximum gain achievable, the parabolic dish.

Corner reflector antennas and slot panel antennas are relatively large devices and, because of this, are not generally made below 140 MHz, to avoid wind loading problems.

Above 140 MHz, the corner reflector and slot panel antennas are a very effective way of achieving higher front to back ratios and predictable coverage patterns.

The other main advantage of using a large screen on an antenna is that the RF effect of the supporting structure on the antenna is minimized, making the antenna particularly suited to applications where a very clearly defined radiation pattern is required, for example cellular radio schemes utilize corner reflector antennas, using a colinear as the driven element in place of a dipole, for sixty degree sectorization.

The corner reflector and slot panel antennas tend to be quite expensive devices, making them more suited to shared user antenna systems where they are more cost effective. They are well suited to multi-user systems such as security forces and public services.

Directional antennas for wide band coverage

The very nature of a yagi antenna construction means the antenna is optimized for a spot frequency and rapidly diminishes in performance moving away from its design frequency. Typically, the effective usable band width is about ±5 per cent of the centre frequency. There are a number of antennas available which will match (give a good VSWR) over a wide band, but normally the rest of the antenna parameters do not follow suit and there is normally a large decrease in gain either side of the design frequency. The log periodic antenna is the only true wide band directional device, but it still limited in the maximum amount of gain it can produce – typically 8–9 dBd. It works differently from other directional antennas in that all the elements are directly driven, not just one as normal. The log periodic consists of a series of half wave dipoles split down two booms. The dipoles decrease in size and the spacing between them gets closer towards the front of the antenna in a logarithmic progression. The antenna can be designed for very large frequency bands, i.e. 68 MHz to 500 MHz, but at any particular frequency only a certain number of elements are active. This gives a very flat frequency response. The antenna is popular with users who need to cover large areas of radio spectrum, such as the broadcast authorities. The log periodic also has the advantage of a high front to back ratio and a very clean radiation pattern, void of any major spurious side lobes.

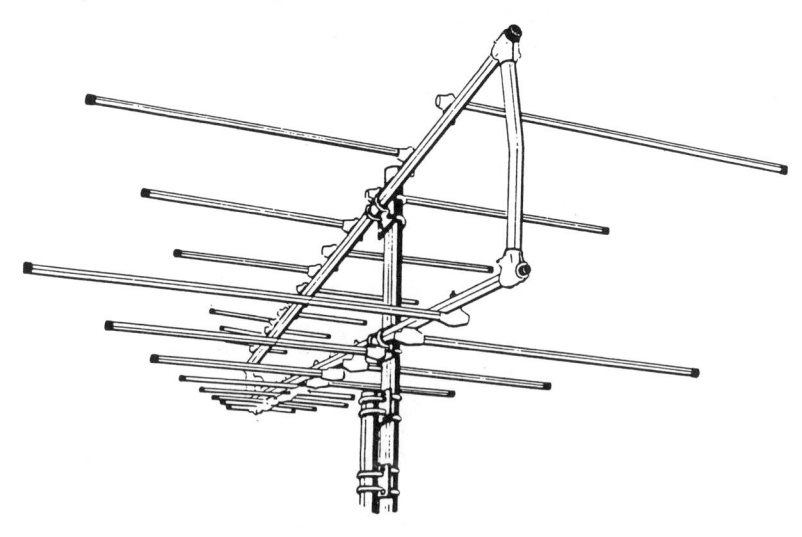

Figure 12.8 *Log periodic antenna at VHF frequency*

Omni directional antenna

The other main group of antennas are those which are designed to give equal radiation in all directions in the horizontal plane and are commonly referred to as omni directional.

Again the simplest form of this type is the half wave dipole. It differs by being fed and mounted from the end instead of the middle, so support booms and feeder cables will not intrude into the radiation pattern.

Although referred to as end fed, the dipoles are normally two quarter wave halves fed in the centre as before but the fed cable is fed down inside one of the dipole arms to exit at the bottom.

The simple end fed dipole can be a metal construction with an insulated section in the centre or it can be made from lighter gauge material and shrouded in a sleeve transparent to radio waves, e.g. glass fibre.

The centre fed dipole is quite often referred to as an omni directional antenna. This is not strictly true as it is designed to be mounted off the side of a supporting structure which will adversely affect its omni directional capabilities.

As with directional antennas, gain cannot be achieved without, at the same time, altering the radiation pattern. As the antennas are still required to give an

Mobile Radio Servicing Handbook

Glass fibre shrouded colinear

Simple end fed dipole

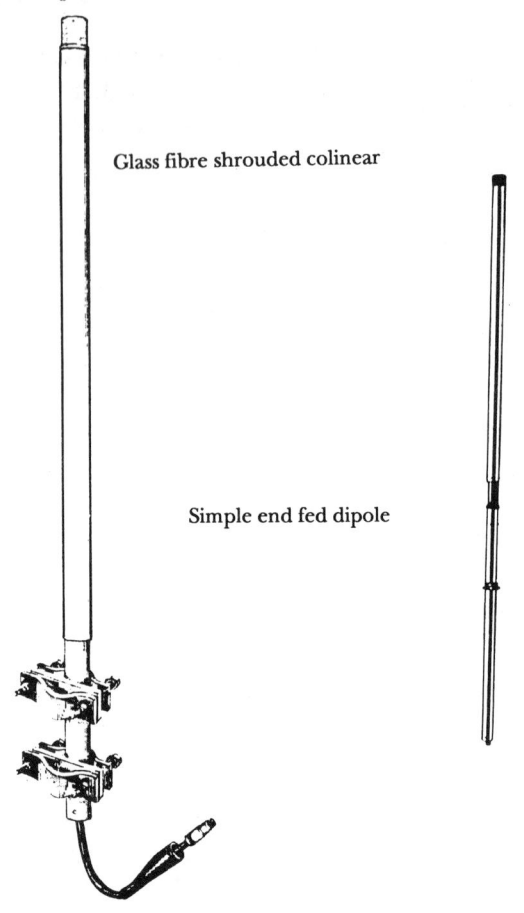

Figure 12.9 *Omni directional antennas, (a) glassfibre shrouded colinear, (b) simple end fed dipole*

omni directional coverage, another method is used to achieve this instead of reflectors and directors. It has already been seen that stacking two identical directional antennas gives an increase in gain of 3 dB. The same method is used with omni directional antennas. Two end fed dipoles are stacked one over the other and joined with an integral combining device. At the correct spacing apart this gives a 3 dB improvement, but at the same time squashes the radiation pattern to give a narrower beam width in the vertical plane whilst maintaining the omni directional pattern in the horizontal plane. Both dipoles are fed in phase with equal power so that they combine. If the number of dipoles is doubled to four, the gain is increased to 6 dB and the vertical beam width is further reduced. This style of antenna is normally called a colinear as the dipoles are stacked in a colinear fashion one over each other. They are normally shrouded in a glass fibre or similar tube.

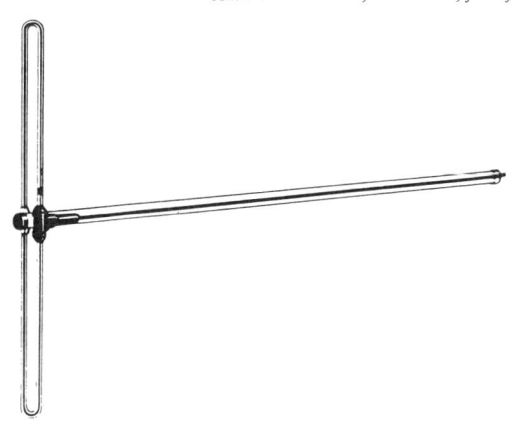

Antennas – selection, installation, fault finding and maintenance

Figure 12.10. *Centre fed folded dipole*

How colinears are fed

There are generally two ways of feeding a colinear antenna. The first is series fed. This is where the power is applied at one end of the antenna, travels through the first dipole then on to the second, usually via a small coil to make sure the dipoles have the correct electrical, as well as mechanical, spacing. The signal then travels through the second dipole, to the third and so on.

The other main method is parallel fed where all the dipoles are fed separately, usually by cable. The cables are then combined together with transforming sections to ensure correct matching.

Both methods give approximately the same overall gain for the same length. The series fed version has a tendency to tilt the main beam because less power is reaching the top dipoles than the bottom. This tends to get worse further away from the design frequency. With the parallel fed version, all dipoles are fed with equal power and do not suffer from the same problems.

As with directional antennas there is a point where increasing the number of dipoles is outweighed by the losses incurred in the device, and the maximum length becomes impractical.

The maximum gain colinears which should be considered are 3 dB at lowband, 6 dB at highband, 9 dB at UHF and higher frequencies.

This method of mounting dipoles in a colinear fashion can be used with centre fed dipoles mounted on a centre support boom. The four stack dipole at UHF is a common antenna using this method. It does, however, suffer from a ragged radiation pattern because of the small number of dipoles it uses to achieve an omni directional pattern. In practice it is much better to use eight dipoles mounted in an 'H' configuration. This gives an improved result – still not as good as the shrouded device.

Mobile Radio Servicing Handbook

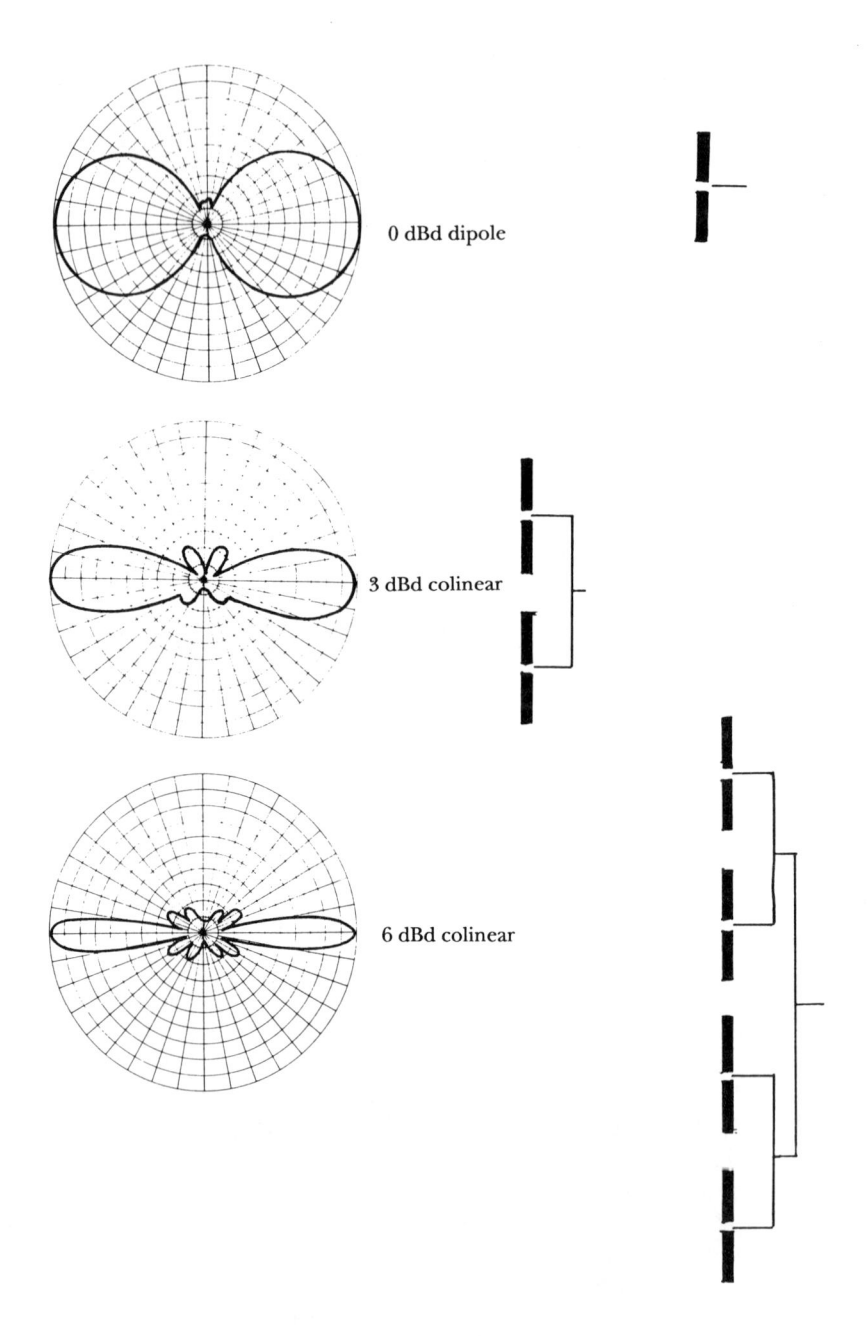

0 dBd dipole

3 dBd colinear

6 dBd colinear

Figure 12.11 *How gain is increased with a narrowing radiation pattern using colinears*

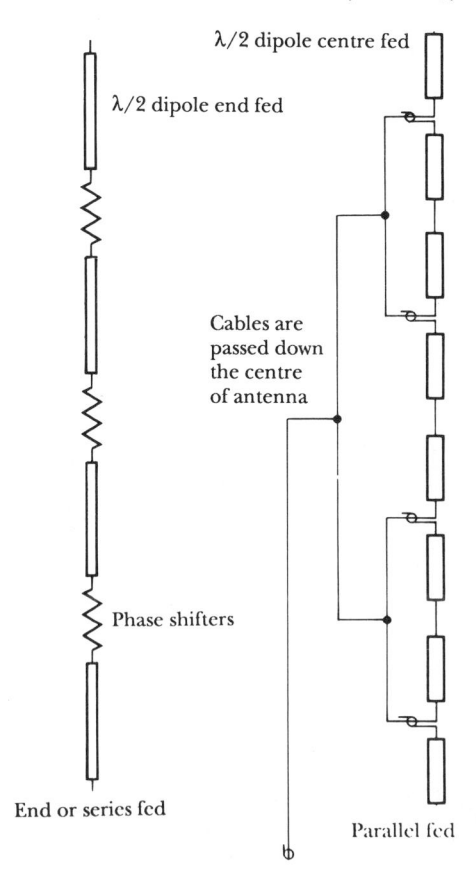

Figure 12.12 *General construction of colinears*

The choice of shrouded colinear or stacked centrefed dipoles is normally influenced by the type of mounting application.

Beam tilt

The main beams on stacked centre fed folded dipoles and parallel fed colinears can be intentionally tilted downwards and upwards. This is normally done to increase the coverage nearer the antenna site, and is achieved by altering the electrical phasing between the dipoles, pulling the main beam down. This can only be done at the time of manufacture – the same antenna cannot be adjusted between tilt and no tilt. Antennas with tilt are, for example, used on cellular schemes where the cell size is small, to avoid signal going over the cell boundary.

Mobile Radio Servicing Handbook

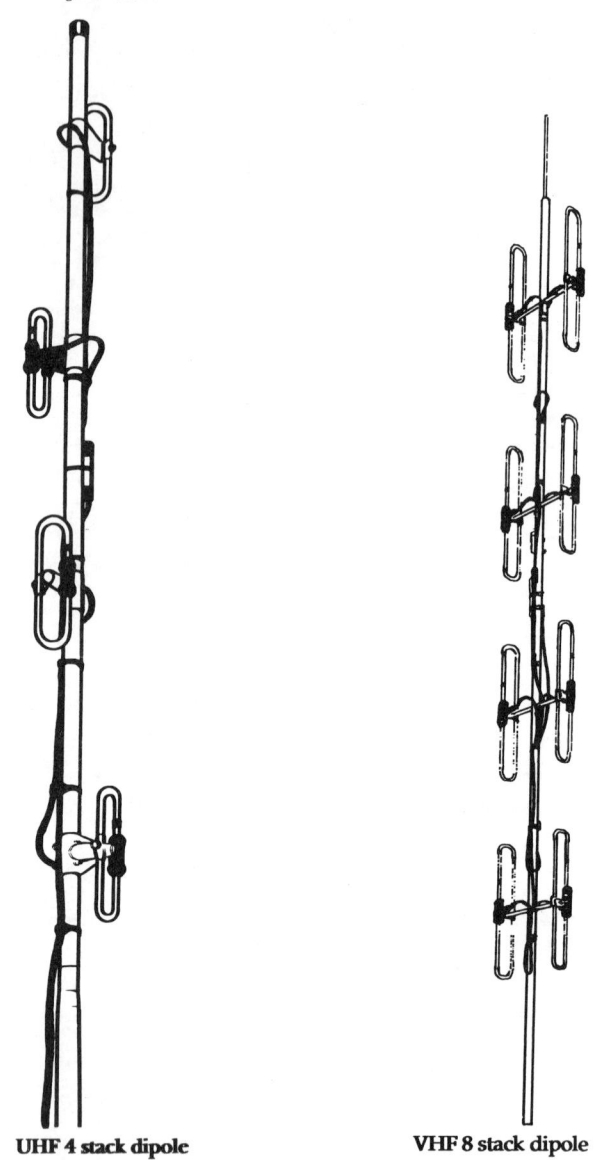

UHF 4 stack dipole **VHF 8 stack dipole**

Figure 12.13 *Stacked dipole arrays*

Baying omni directional antennas

Phasing together two omni directional antennas mounted side by side and varying the phasing and mechanical spacing between them can generate a variety of radiation patterns. A common one is with a ⅜ wave length spacing between two end fed dipoles fed in phase. This produces a long thin pattern, ideal for a central base station to service a narrow corridor of coverage.

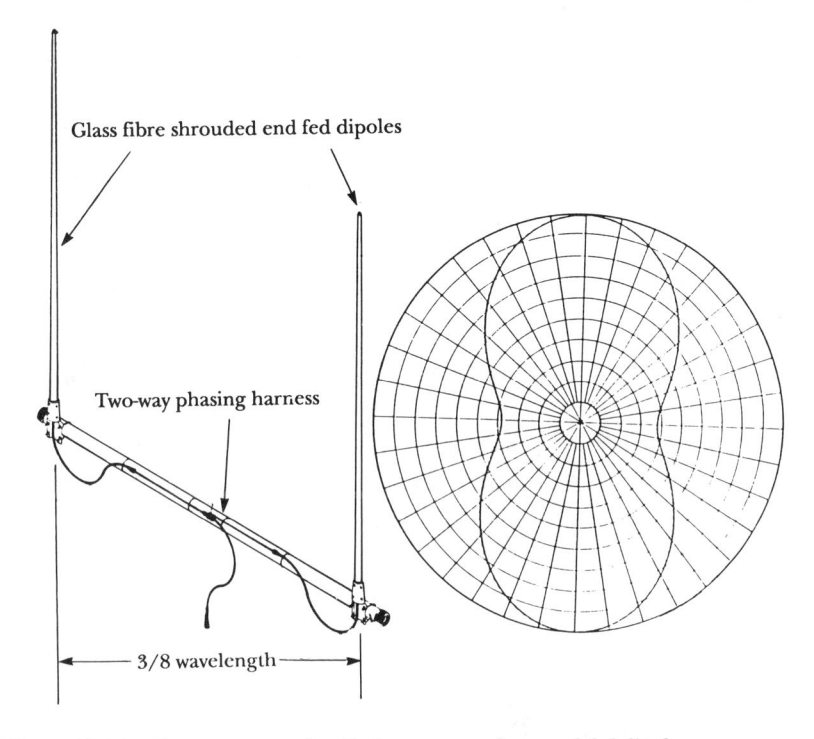

Figure 12.14 *Arrangement and radiation pattern of two end fed dipoles*

Cardioid dipole

This is another common form of phasing two antennas together, usually with a pair of centre fed dipoles either on a common boom or one above each other. The dipoles are spaced a quarter wave length apart with the dipole furthest forward having a quarter wave longer tail to the phasing harness. Signals received at the front of the pair are in phase when combined, signals received at the rear are 180 degrees out of phase, giving a very high front to back ratio. This principle can be applied to any pair of directional antennas and is normally referred to as quadrature phasing. It is also in common use with two and four stacked dipoles to give a guaranteed high front to back figure. These styles of antenna are being used to fulfill the 20 dB front to back requirement on the Band III trunked systems.

Mobile Radio Servicing Handbook

Cardioid dipoles

Four stack dipole fed in quadrature

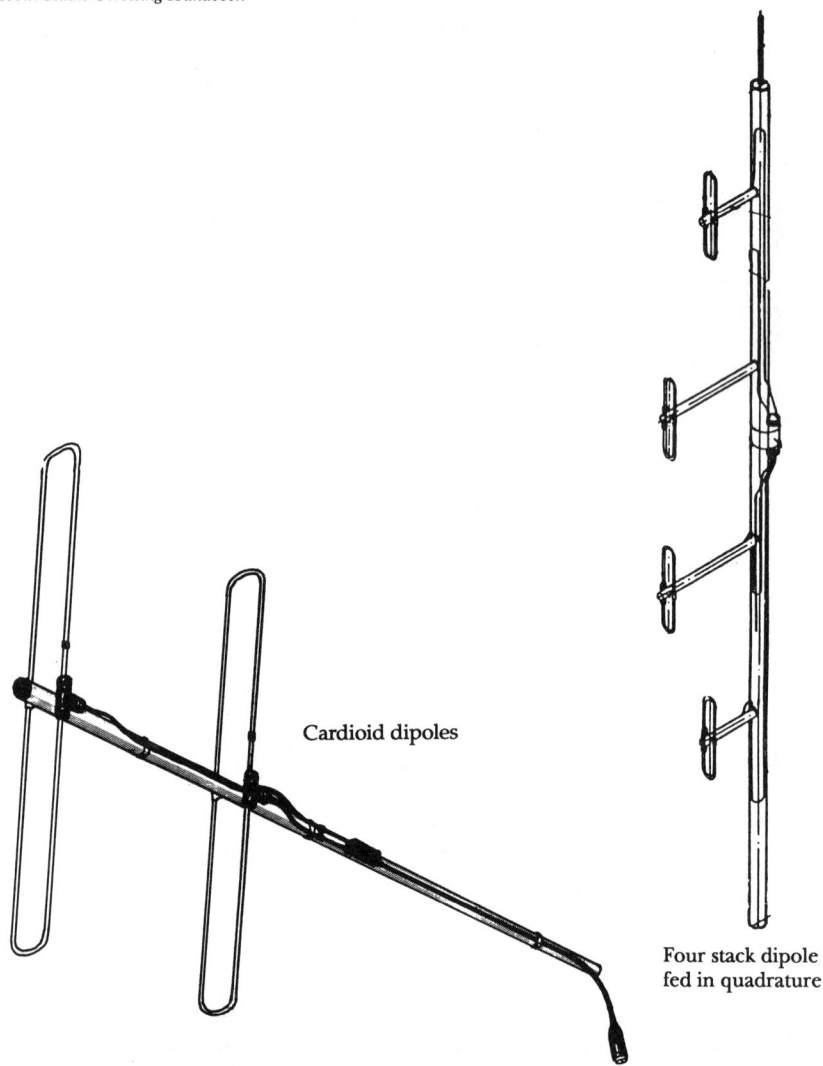

Figure 12.15 *Various forms of cardioid dipoles and quadrature feeding*

Choosing omni directional antennas

All the omni directional antennas which have been referred to are vertically polarized. It is difficult and expensive to produce omni directional antennas which are horizontally polarized. This is not the case for directional antennas where turning the antenna through ninety degrees will change polarization.

Until recently, not having simple horizontally polarized omni directional

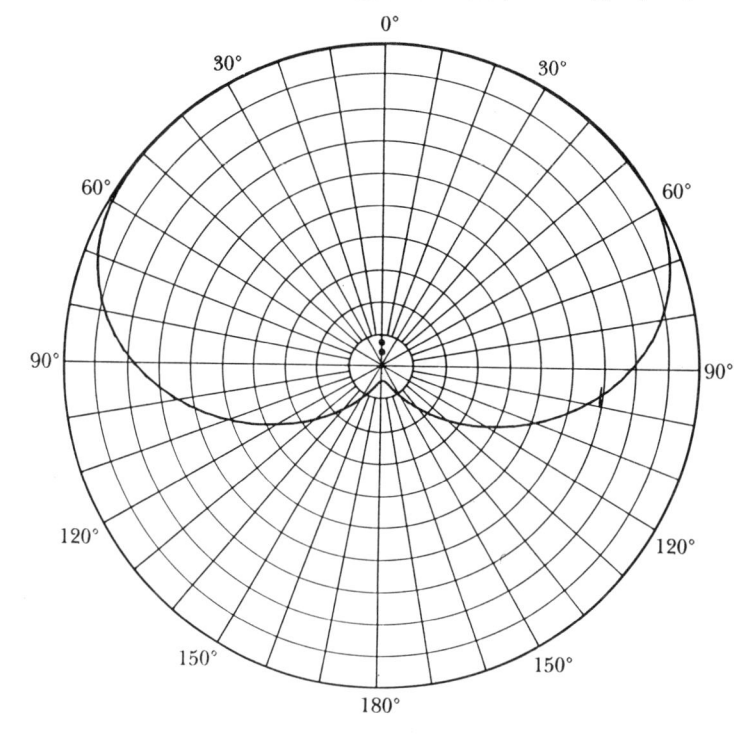

Figure 12.16 *Typical radiation pattern for cardioid dipoles (quadrature feeding two half-wave dipoles)*

antennas has not mattered as there has been no real requirement. However, some specifications such as the one for telemetry (MPT 1411), specify this option to improve use of the available spectrum. As the demand is still limited and the manufacture of these antennas is difficult, they tend to be expensive.

The polo mint effect

As has been seen, the doughnut shape of the radiation pattern is squashed out when achieving gain from an omni directional antenna. This tends to create a hole in the centre of the radiation pattern which increases in size as the gain increases. This can cause a significant null in coverage close into the antenna. For example, a 3 dB colinear will provide, say, twenty-seven miles of coverage at 25 W ERP with a 400 ft 'hole' at the centre. A 10 dB colinear at the same height and frequency will give thirty-four miles of coverage, but with a 2400 ft hole. The higher the mounting height the greater the problem. The user will be very dependent on reflections from the supporting structure and spurious side lobes to achieve close in coverage. Improving the distance covered from base

Mobile Radio Servicing Handbook

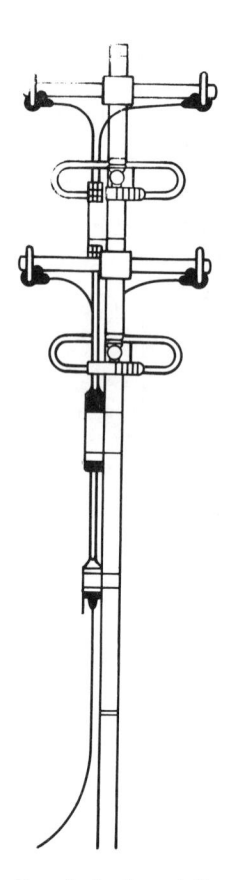

Figure 12.17 *Typical horizontally polarized omni directional antenna*

site could result in the close in coverage being greatly reduced. Note that all antennas, directional and omni directional, will have a null shown on the typical radiation patterns. When taking field strength measurements, for example to establish maximum ERP (effective radiated power), a measurement should always be taken some distance from the supporting structure.

Wide band omni directional antennas

The only wide band omni directional antenna is the conical or biconical antenna. The conical is the most common form with small top hat and wide skirt. They have a unity gain and are designed to cover a band width of 3:1, i.e. 100–300 MHz. They do not generally have a good VSWR figure, and are best used for general purpose receive applications or as a standby for several antennas within its frequency band.

Note, these antennas are often unrealistically expected to cover very large frequency bands, i.e. 100–500 MHz and the result is poor performance.

Various other styles and applications for antennas

Circular polarization

This is either in the form of helical or cross yagis (two identical chains of elements mounted at ninety degrees to each other on a common boom). It is used for long distance communication where there is a chance of the signal being twisted during travel, and also for satellite communication where the satellite is spinning, producing a circular signal, it also used to minimize the effect of multi-path propagation losses occurring in over water transmissions. Antennas can be built for signals spinning clockwise or anti-clockwise. Care must be taken if mixing linear and circular polarization as a loss of 3 dB will occur as half the signal is lost in the opposite plane.

Another form of circular polarization is used by the broadcast authorities for FM radio transmission. This is normally referred to as mixed polarization. The listener's antenna can be vertically or horizontally polarized.

Paging antennas

A form of antenna with mixed polarization designed to radiate downwards, with very localized coverage, for example coverage for a radio paging system within a building where the antenna sits on top of the building.

Mobile and portable radio antennas

The performance characteristics (expressed in decibel gain) of mobile and portable antennas are commonly and erroneously compared with the characteristics of fixed base station antennas; there is no direct comparison

Vehicle mobile antennas

The most common form of vehicle mobile antenna is the quarter wave length whip antenna. It must be remembered that all mobile antennas are in fact only half what is required to make a true efficient antenna, they all require a ground plane or counterpoise, and therefore rely on the vehicle body work to supply this ground plane. The reception/transmission characteristics of mobile antennas are mainly dictated by the position of the antenna on the vehicle body, so careful siting is required. The best position, found in testing, is the centre of the vehicle roof. Moving towards the front or rear of the roof degrades performance, but not to a great extent. Placing the antenna on vehicle wings or boot lids produces the worst performance as the vehicle body protruding above the antenna changes the radiation pattern, creating dead spots when the vehicle is in certain directions relative to the base station antenna site.

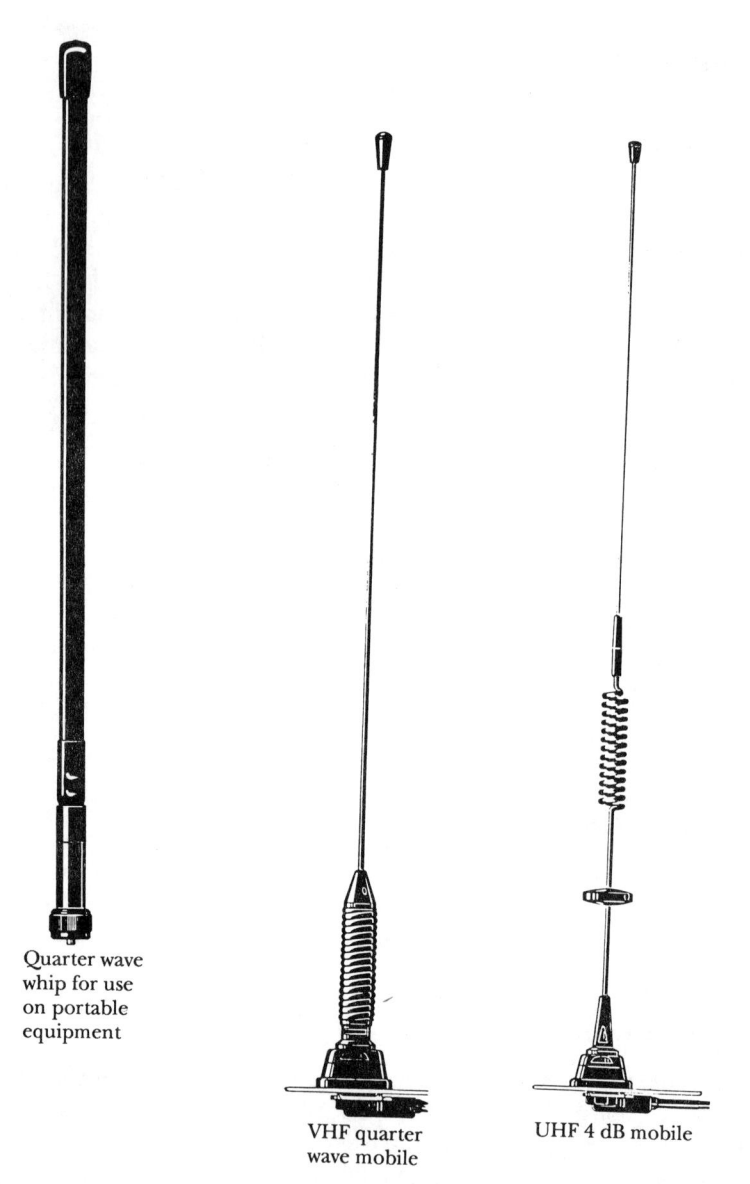

Quarter wave
whip for use
on portable
equipment

VHF quarter
wave mobile

UHF 4 dB mobile

Figure 12.18 *Antennas for mobile and portable use*

Performance is also adversely effected by poor earthing of the antenna to its ground plane, i.e. the vehicle body.

Important points to remember are:

1 The need for a good ground plane. Glass fibre roofs, for example, must have a metal or foil plate to compensate beneath the antenna mount. This can be located in the vehicle between the roof lining and the roof.

2 A good earth between antenna and ground plane.
3 Siting, middle of roof for best performance.

Diversity reception using two antennas per car may become a practical alternative and may offer worthwhile reception/transmission benefits with data over radio transmissions in particular.

Portable antennas

As with mobile vehicle antennas, portable antennas designed for radios carried in the hand or on the body, tend to be quarter wave length whips. These suffer from being in close proximity to a solid mass which can shield the antenna and greatly upset the impedance, degrading performance. The small size of portable radios tends to give a minimal ground plane for the antenna to work from. Due to the very nature of this type of equipment, little can be done to improve performance apart from making operators aware of the problems, to avoid having the radio too close, and from standing between the equipment and the site to be reached. There are a few antennas recently available which offer an integral ground plane with the antenna, thus making it a true end fed dipole. This means that they do not rely on the small radio for a ground plane.

Specification of fixed antennas

We have suggested general guidelines for selection of the right antenna for the right application. At the specification stage it helps to have an appreciation of how an antenna is specified technically.

VSWR (voltage standing wave ratio)

The ratio of power applied to the antenna to that reflected back down. A typical VSWR of 1.5:1 is common, showing the antenna has dissipated 96 per cent of the power applied to it. The majority of transceivers are designed to cope with a VSWR of this order. Certain antennas, in particular very wide band devices, might quote a VSWR of 2.0:1 or greater, check to see if the equipment on which it is being used will work with this higher figure.

Maximum power

The amount of power an antenna can handle continuously, check if the figure quoted is PEP or CW.

Gain

There are normally two ways of quoting gain. The most popular and meaningful figure is dBd – gain relative to a half wave dipole. Occasionally a gain figure is quoted in dBi – gain relative to an isotropic radiator. This isotropic radiator is a theoretical antenna which radiates equally in all directions and the 3D radiation pattern takes the form of a sphere. As it is a theoretical device it makes more sense to use the practical dBd. Check which figure is quoted; if dBi is quoted subtract 2.15 dB to get the dBd figure (dBi − 2.15 = dBd).

Half power beam widths

A figure quoted as a guide to show how wide the main lobe of an antenna's radiation pattern is. It is measured in degrees across the main beam where it has reached half power, 3 dB from peak gain. Occasionally 6 dB points are quoted as well.

Band width

The band of frequency spectrum across which an antenna has been designed to perform at its best in all parameters. Normally quoted as a percentage of centre frequency.

Cross polar discrimination

A measurement of how well an antenna can discriminate between the correct polarization when receiving a linear signal, and the opposite incorrect polarization. A typical figure would be 20 dB. This measurement has become much more relevant with the new telemetry specifications, where licences are being issued with either vertical or horizontal polarization to maximize the available frequency spectrum. An antenna must reject unwanted signals in the opposite plane.

Front to back ratio

A measurement in decibels of the loss on a received signal when received off the rear of an antenna, as compared to the front.

Input impedance

The design impedance of the antenna, normally 50 ohms. Check to see if the radio equipment accepts this impedance.

Connection

Which connector is required and are there any special lengths of antenna cable needed? Manufacturers tend to have a standard tail length and connector which they offer unless asked for an alternative.

Weights and windloadings

Important figures for gauging the additional loading which the antenna will supply to its supporting structure. Check also if the antenna will withstand the anticipated weather conditions.

Interpreting polar diagrams or radiation patterns

Polar diagrams shown in manufacturers' catalogues are taken with the antenna in free space, i.e. with no interfering metal work around it. They are primarily a visual design aid to show the best coverage which an antenna will give. They should be used in conjunction with propagation prediction, taking into account the amount of signal being radiated. They are a gauge of how far a signal will radiate from a particular antenna – a diagram of the direction in which signals will be radiated at relative levels.

VSWR and gain graphs

A guide to the performance of the antenna in gain and match over its design band. Gain graphs normally remain accurate when applied to the field after being measured on a test site. VSWR graphs however, do tend to vary due to mounting differences. The VSWR of the antenna should not, however, exceed the quoted maximum.

Assessing an antenna's performance

Almost alone among the various components that make up a radio system, testing an antenna for all its major parameters is beyond the scope of normal radio workshop test equipment.

The only test open to users is doing a VSWR measurement using an in-line power meter which can look at forward and reverse power. This can only show how much power is being dissipated by the antenna and does not even show how much of that power is actually being transmitted to the atmosphere and how much is lost in the antenna. One has to rely on a manufacturer's integrity in writing specification sheets for the rest of an antenna's performance.

The manufacturer will assess an antenna by performing actual measured tests on a known test range in the best possible conditions. This takes the form of radiation patterns in both planes, radiation patterns in cross polar form and measured gain checks which look at the gain of the complete antenna and not just computed results. This will show if any power is being lost internally.

These results are then used to write the catalogue specification sheets.

Similar antennas from different manufacturers should have similar results, so it is worth comparing catalogues. It should, however, be remembered that an antenna's performance is directly related to its overall size (its aperture). This means that for a certain gain figure an antenna must be at least a certain length long. If the antenna is shorter it will not have enough gain. So if one manufacturer claims gain X for a particular length, and another manufacturer also claims gain X for a similar antenna but in a much shorter length, one of them is wrong. There are no laid down standards for producing antennas to the same electrical specifications apart from the more general MPT specification from the DTI. Reliable manufacturers will be able to back up their claims with more detailed patterns or gain charts if requested, and would be happy to show customers their testing procedures. A word of warning – if radiation patterns and gain curves are wanted for the specific antenna ordered, the manufacturer is likely to charge for this if his standard graphs are not acceptable.

Choice of radio site

These days it is harder to create a new radio site, due to environmental constraints. It is more than likely that an existing radio site will be available. The choice of existing radio site will be the one which gives the coverage required, found by consulting with site owners.

If a new site is needed then it is essential that proper propagation analysis is carried out to ensure the coverage required is obtainable. It is also essential that propagation analysis undertaken by theoretical means, using one of the many computer services available, is backed up by physical studies. This is because the data base used in computer studies cannot take account of unexpected changes, such as new buildings being put up or old ones demolished.

One other major consideration when choosing an area to put a new radio site is the level of potential interference. Noise can be generated from a close busy road, high voltage power lines or similar, so avoiding possible sources can help stop annoying noise levels intruding on communications.

Finally when appraising a possible site, attention must be paid to availability of electrical power and telephone lines. Care should also be taken over how the final structure, complete with antennas and equipment buildings will look. They should be made as unobtrusive as possible to avoid potential complaints.

The job of planning a new radio scheme is made simpler by using an existing site. When approaching an existing site the choice will be between:

1 Sharing a combined antenna system. This is where the site owner has installed an antenna array for a particular frequency band and puts a number of users on to it, using combining and filtering techniques. All that has to be done is to plug the equipment into the cable tail provided.

2 Putting up a new antenna array and sharing the equipment room, if there is space, or providing a new equipment enclosure if there is not.

The decision on which method to choose, apart from finance, depends on whether the existing antenna array will give the coverage required. The site owner will answer this as it will already be mapped. Will the structure have space for another antenna array and will it take the additional wind loading? Again the site owner or manager will answer this if told precisely which antennas are proposed.

As with a new site, electrical power and telephone line requirements must be investigated.

Whichever method is chosen to implement a new radio scheme, remember prior planning prevents poor performance.

A guide to good radio site management practice is contained in the DTI publication MPT 1331.

Installation of fixed antennas

Poor radio communication can often be blamed on the wrong choice of antennas, incorrect siting of antennas and/or bad installation. A correctly installed antenna system that is properly maintained will last at least ten to fifteen years. A poorly installed system might only last until the first time it rains.

Choosing the right antenna for the job has been discussed. Various problems can, and do, occur in installing antennas.

Mounting antennas off the side of structures

Mounting any antenna off the side of a structure will induce currents into the structure. This can easily cause intermodulation products and interference. However, directional antennas such as yagis and those with good front to back ratios, generally do not cause a problem with the antenna pointed directly away from a structure. It is recommended, where practicably possible, to have a minimum spacing of one wave length between the structure and the reflector of the antenna.

It is of course a different matter for omni directional antennas such as end feds and centre fed folded dipoles. Omni directional antennas should be placed on the top of a supporting structure, clear of any surrounding obstructions, in order to achieve an omni pattern. This prime position is normally the first to be filled. The correct alternative solution would be to design a multiple antenna system around a structure which, when combined, would give a reasonable omni directional pattern. For example a square tower with a face width of 1.2 m, and a centre frequency requirement of 200 MHz, would have a solution of four slot panel antennas arrayed round the tower, one on each face at the same level and combined into a single down lead with a four way phasing harness. The resulting radiation pattern is shown in Figure 12.19.

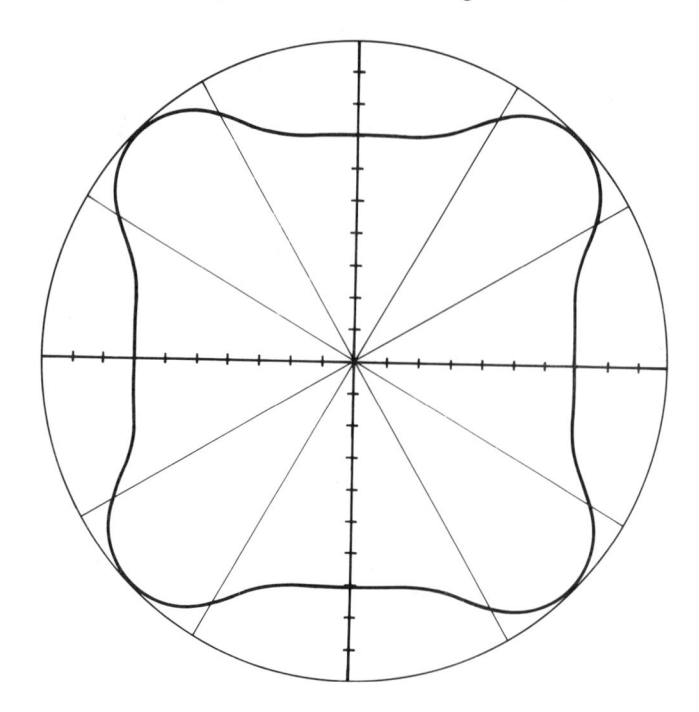

Figure 12.19 *Radiation pattern of four slot panel antennas placed one on each face on a 1.2m square tower*

This is typical of the pattern which should be aimed for when combining antennas for an omni directional pattern, with the troughs in the pattern being not more than 3 dB less than the peak gain. It is, however, difficult and expensive to achieve. The planning of such a system is best left to antenna designers/manufacturers. Being expensive, it is normally only viable for multi user systems.

It has become accepted practice, although not strictly correct practice, to mount end fed dipoles, centre fed dipoles and colinears, off the side of structures. Two potential problems occur:

1 The antenna is radiating directly into the structure, thus being the possible cause of interference problems.
2 The structure will greatly alter the radiation pattern characteristics of the antenna and, if it is too close, the VSWR characteristics as well.

To minimize this, omni directional antennas should be spaced not less than two wavelengths from the structure and checks should be made after installation on how much the radiation pattern has been altered and if there is any interference being caused to other site users.

Mounting materials

All brackets, clamps and off mounting steelwork should be made of a material compatible with the antenna and the structure. Putting two dissimilar metals together causes an electrolytic action, as in a battery. For example, copper and aluminium do not mix and if placed together the aluminium will eventually be eaten away.

The choice of supporting bracketry is common sense, but some site owners will have preferred methods and materials. Generally the preferred materials are aluminium, galvanized steel or stainless steel.

Coaxial cables and connectors

There are two main types of cable which are in common use with antenna installations.

1 Solid dielectric cable

Most commonly a stranded copper inner with a polyethyline dielectric, a braided outer covered with a plastic jacket, e.g. RG213/U (URM 67). At higher frequencies or when used with combiner/filter systems, a double screened cable should be employed, e.g. RG 214/U.

Advantages	Disadvantages
Low cost	High loss
Easy to install	RF leakage
Flexible	Short life

2 Foam dielectric cable

A solid copper or copper plated aluminium inner with foam plastic dielectric, a plain or corrugated (for strength) outer made from copper with a plastic jacket. Common sizes are ⅜in, ½in or ⅞in diameter.

Advantages	Disadvantages
Low loss	Not very flexible
No RF leakage	High cost for cable
Long life	and connectors

The choice of which cable type and size to use depends on:
- Frequency (loss increases with frequency)
- Length of run
- Weather conditions
- Mounting position
- Cost

Always follow the cable manufacturer's recommendations on how far apart to place cable support cleats. If cleats are too far apart, the cable may move with the wind, eventually causing damage. In particular the foam filled cables with solid outers can fracture, which shows up as an intermittent fault, difficult to trace. There are a number of cable support systems available which will support the cable without crushing it, site owners may have a preference. The one golden rule when planning a feeder run is to avoid places where it may be stood on, always put cables under structure cross members and never run cables down an access ladder.

Connectors

There are various types of connector systems in use, the main one being the N type which has overtaken the UHF series in popularity. The N type is a much better style of connector than the UHF series, with better loss figures and better matching. The choice of connector is not critical as long as the connectors chosen are of a good quality. Using UHF series connectors above 300 MHz is not recommended.

Whichever connector series is chosen, it is very important that they are correctly waterproofed. The best method is as follows: wrap the joint with self-amalgamating tape going at least two inches either side of the connectors, then cover with PVC overwrapping tape. The joint is now waterproof, but it is recommended that a third layer of Denso tape, or similar, is put on top. This gives a joint which will never let in water. If it has to be opened it will be clean inside.

Isolation

An important point on site sharing is the problem of intermodulation products. These are caused by the mixing of two or more frequencies producing other unwanted signals. These can be caused by non-linear mechanical junctions on supporting structures and antennas, by dissimilar metals and by inadequate isolation between antennas. An isolation figure of about 40–45 dB between a transmit antenna and a receive antenna, and a figure of about 20–25 dB between two transmit antennas.

Post commissioning inspection

Remember, many problems with poor coverage, low gain, bad cross polar response, intermodulation and poor VSWR, are caused by incorrectly fitting an antenna.

Antennas – selection, installation, fault finding and maintenance

Positioning of antennas on structures – the major points

1 Always point directional antennas away from a structure, not alongside, as this will induce currents into the structure and affect radiation patterns.
2 Never introduce supporting metal work into the antenna chain in the same plane. The antenna cannot distinguish between a support pole and an element, and this will greatly affect the radiation pattern and the VSWR. If metal work has to go through the element chain, fix it at ninety degrees to the elements, or use materials transparent to radio waves if it cannot be avoided.
3 Attempt to mount omni directional antennas on top of supporting structures, or failing that, provide a minimum spacing of two wavelengths between antenna and structure.
4 Choose mounting bracketry which is made from materials compatible with both structure and antenna.
5 Ensure there is enough separation between the antenna and existing antennas on a structure, in particular those on similar frequencies.
6 Choose a feeder cable which has a signal loss through it for which the radio equipment can compensate.
7 Install feeder cable in accordance with the manufacturer's recommendation.
8 Properly waterproof all joints.
9 Ensure antenna and feeder are marked for easy reference and are correctly earthed to ground.
10 Inspect regularly.

Following the above recommendations should avoid a call out on a wet and windy night to rectify an antenna problem.

Fault diagnosis

The majority of faults on antenna systems usually show as reduced system coverage and on inspection as a high VSWR. Checking the antenna for VSWR at the bottom of the feeder run will not show whether antenna, cable or connectors are at fault. Antennas and feeders should be tested separately, especially as a long feeder run can hide a faulty antenna. Both antennas and feeders can be tested for VSWR and DC short. For cables, a load or short circuit will have to be applied at one end. The majority of antennas will show a DC short. This is normal as it is there to provide protection from static interference.

It is recommended that a record is kept of VSWR measurements from installation, with a reading taken and recorded when the radio equipment is commissioned. Problems such as water ingress tend to show as a gradual deterioration in VSWR, so if a record is kept there is every chance of spotting the problem before the system fails.

Environmental problems

Corrosion

Corrosion can occur in two forms. The first, already mentioned, being dissimilar metals. The second is that untreated materials, such as steel, will rust. This in turn will provide bad contacts, in particular on mounting brackets which can give rise to unwanted intermodulation products.

Weather

Wind, rain, snow and ice are an antenna system's worst enemy, after a rigger. Wind can cause vibration in antennas leading to metal fatigue. This is usually solved by careful mounting and the use of damping material inside elements if the problem is too bad. 6 mm terylene rope along the inside of an antenna element works wonders. The wind will also make short work of badly fitted cables, damaging them and allowing further problems of water ingress.

Ice and snow build up on an antenna will have the short term effect of creating poor VSWR and reducing coverage. The additional weight or ice falls may break the antenna. Ice will also split cables and antenna elements if water, allowed to enter through damage, freezes.

If icing is a particular problem on a site, and it is important that radio circuits are kept open, antennas can be supplied with heated elements to stop ice formation. The low voltage heaters are supplied by a transformer in the equipment room with an external frost stat. Heating is expensive and only recommended for problem sites.

Pollution

Heavy industrial pollution or sea spray can attack the aluminium construction of antennas. If it is a particular problem, the antenna can be coated with various finishes, e.g. polyester coating, nylon coating or PTFE coating. Again these coatings will add to the cost of the antenna system, but should only be required where pollution is particularly severe.

Health and safety

Lightning protection

The power of a typical lightning strike equates to several kilovolts per metre.

Antennas – selection, installation, fault finding and maintenance

The need for good lightning protection is firstly to protect life, and secondly to protect equipment. The recommended way of grounding equipment is shown in Figure 12.20. The major points to remember are:

1 A good common ground joining all equipments and masts.
2 Provide the lowest inductance paths possible from all equipment to ground.
3 Make all earth connections as short as possible with no sharp bends.

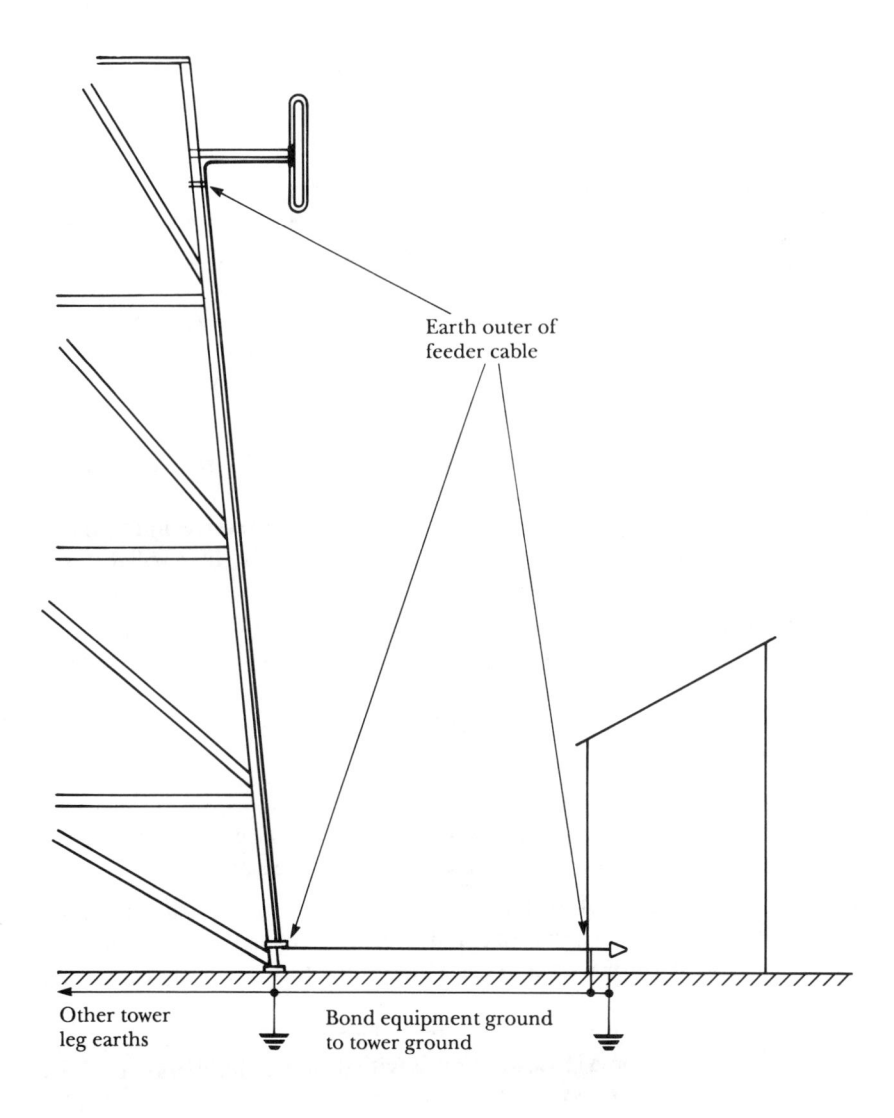

Figure 12.20 *Recommended grounding of antennas to towers*

Radiation hazard

This is a subject on which much research is continually being done and it is advisable to watch for recommendations from the National Radiological Protection Board and the DTI.

The current UK limit is 100 W/MTR^2. It is very easy to find these levels of radiation on radio structures, especially on multi-user sites.

If a radio structure has to be climbed, for personal protection, check the sort of signal levels which are on it. Some antenna arrays may be unsafe to climb past or work near without reducing power or switching off.

Duplexers and filters

It has already been seen that for radio systems to work well there must be a degree of isolation between antennas on the same system. Without this isolation, receivers would become desensitized by unwanted signals from nearby transmitters swamping the front end of the receiver.

There are two ways of achieving this isolation. One is to separate the antennas by the correct distance, as already mentioned, and the other is to use a duplexer. The disadvantages of physically separating antennas are:

- Availability of space on a supporting structure. Obviously it is easier to find space for a single antenna and if the space has to be rented it is less expensive.
- Radiation pattern. When using a system with separate receive and transmit antennas it is highly likely that the radiation pattern from them will differ as they are in different positions on the structure.

Isolation between transmit frequency and receive frequency can be difficult to predict due to reflections which occur from the supporting structure, and the likelihood of interference from other users' systems.

Thus the advantages of using a duplexer to a single antenna instead of separate antennas for transmit and receive are:

- the duplexer will always provide the correct isolation;
- only one antenna, one feeder run and one antenna space are required;
- the radiation pattern for transmit and receive are the same.

The main disadvantage of using duplexers is that they introduce some additional losses into the system. Also, if different antennas are needed on each side, for example to improve the receive side, then this obviously cannot be done.

There are two basic types of duplex filters; the *band reject* and the *band pass* duplexers. Both types come in various forms with different numbers of cavities and can either be built up from individual cavities or come as a 'ready made' unit.

Band reject duplexer

This style of duplexer is probably the most common type in service as it combines compactness with good performance. The band reject duplexer is made up from a number of band reject filters, typically three filters in each half, i.e. the transmit side and the receive side. The band reject filter has the ability to attenuate a narrow band of frequencies while allowing other frequencies to pass through with very little loss. Maximum attenuation occurs at the resonant (design) frequency and sharply decreases as you move away from this point. The signal at the resonant frequency sees the filter as a block and is lost within it. The filter will provide this attenuation no matter how close or far apart the pass and reject frequencies are. The only factor in how close the two frequencies are is how much signal loss is acceptable in the pass band. The reject curve is, however, very steep, which allows very close spacing. The reject curve can be given various profiles by the use of additional coaxial stubs, i.e. a symmetrical curve (both sides of the curve are the same); a high pass curve where the higher frequency side of the reject curve is much steeper; the low pass curve where the low frequency side is much steeper.

The filters can be connected to each other giving a much larger attenuation, typically two filters give about twice as much attenuation as one.

When used in the duplexer, a number of filters are interconnected in a 'duplex' fashion with transmit on one side, receive on the other and the antenna in the middle. The filters on the transmit side are tuned to reject the receive frequency and the filters on the receive side are tuned to reject the transmit, thus received signals will be accepted down the receive leg but any unwanted transmit signals will be attenuated, protecting the receiver.

This style of duplexer has very low insertion loss and very good rejection. The duplexer can also be made using helically wound cavity filters, making it much more compact so it can become an integral part of the radio equipment.

Band pass duplexer

The band pass duplexer is made from band pass filters. The band pass filter is very common but the band pass duplexer is not used as much as its band reject rival. The band pass filter is a device which can let through a narrow band of signal while all signals outside the band are attenuated.

Adding more filters will improve the attenuation of the device, not forgetting that the insertion loss will also increase slightly. The band pass duplexer is made up from a number of filters (cavities), those in the receive leg are tuned to the receive frequency and those in the transmit leg to the transmit frequency. As each leg will only pass or accept its required frequency signal and attenuate all other signals, both legs of the duplexer are protecting the equipment.

As this style of duplexer only passes the frequencies wanted for the particular piece of equipment to which it is attached, it can be seen that it is very 'other user' friendly. On the transmit side it only transmits the narrow band of wanted signal and attenuates all other unwanted signals such as harmonics, reducing the amount of noise on the radio site. Also on the receive side, only accepting the receive wanted signal can help to reduce interference. However, the band pass duplexer, although being a very simple and useful device, is not suitable for narrow frequency spacings as the maximum attenuation occurs some distance from the design frequency, restricting this type of duplexing to wide frequency spacings.

Combining

Both the above styles of duplexer can be used not only just for the receive and transmit frequencies of a single radio, but also to combine together two receivers, two transmitters or two simplex radios. They are often called diplexers or combiners when used in this configuration. It should not be forgotten that a duplexer is only a device to provide isolation between two separate frequencies within, of course, the design range of the diplexer chosen. It should also be remembered when combining two transmitters that the *combined* power output of the transmitters does not exceed the power handling capacity of the duplexer.

Various other forms of combiners exist, such as the receiver multicoupler. This is a device to split the received signal from a single antenna into any number of separate receivers. They normally employ an amplifier, boosting the signal to the receivers to compensate for the splitter losses.

Note that the amplifer is only there to boost the received signal up to an acceptable level prior to splitting.

The ring hybrid is a common method for connecting a number of similar frequency transmitters together into a single antenna array, most commonly two or four transmitters. The ring hybrid works on the simple principle of out of phase signals cancelling. By the means of quarter wave length multiples within the coaxial ring, the transmit ports are kept 180 degrees out of phase with each other, directing the power into the antenna port. The ring hybrid is a simple device but will also need circulators on the antenna ports to ensure no signal is allowed back down the transmit ports. All these components can add up to a high insertion loss within the device, typically around 7.5–8 dB for a four transmit hybrid. The other disadvantage is that as the hybrid relies on lengths of coaxial cable cut to precise frequencies, the band width is small.

Specifying duplexers

There are very few parameters to note on most duplexers as they tend to become standard items for the common frequency bands. The question most usually asked is, how many cavities? The more cavities the greater the rejection figure, however the more cavities, the greater the insertion loss. So referring to the manufacturer's data sheets is important.

Antennas – selection, installation, fault finding and maintenance

- Insertion loss is a measurement of how much of the wanted signal is lost through the device.
- Rejection or isolation is a measurement of how much the unwanted signal is attenuated. This figure will be shown for both transmit and receive legs.
- Power handling is making sure the device will cope with the size of transmitter to which it is coupled, especially important when combining two transmitters.
- Separation is the figure given in MHz to show how far apart the two frequencies must be for the duplexer to perform correctly. This can vary or be fixed.
- Temperature range; as duplexers/filters will change their characteristics with major changes in temperature, it is important to try to keep the temperature fairly constant, and obviously choose a device which will handle varying temperatures if this cannot be controlled.

One final point with filters and duplexers is that they should not replace the proper spacing of antennas on their supporting structure. Even if the filter/duplexer is providing the correct isolation for your system, it will not be helpful to other site users.

Dish antennas for microwave and millimetric wavelengths

Dish antennas for microwave and millimetric wavelength applications – for line of sight ground links and satellite links – are a topic in their own right, but some general comments related to land mobile radio applications are appropriate.

As wavelength decreases, smaller more compact antennas become possible and higher gain is achievable, off-setting to some extent the higher attenuation characteristics at microwave/millimetric frequencies.

Gain of a dish antenna can be expressed as: Gain = $(\pi D/\lambda)^2 \times N$ in which:

D = Diameter
λ = Wavelength
n = Efficiency

The limits of gain are ultimately set by the tolerance of the reflecting surface. 30–50 dBi is achievable with 1° beamwidths.

This gain needs to be compared with the attenuation over, for example, a 20 km link of

118 dB at 1 GHz
132 dB at 5 GHz
138 dB at 10 GHz

Surface irregularities set a limit to the ratio of the focal distance (the distance

Mobile Radio Servicing Handbook

between the feed point and the centre of the dish), to the reflector diameter – known as the f/D ratio. Practical values are between 0.2 and 1. The higher the value, the better (electrically) and more difficult (mechanically).

Practical problems, such as susceptibility to damage, occur if the feed point extends beyond the front of the rim.

Reducing the focal distance to avoid this can be achieved by using two alternative forms of dish antenna.

- *The Cassegrain* – which uses a convex hyperboloid sub-reflector.
- *The Gregorian* – which uses a concave ellipsoidal sub-reflector.

In both cases the disadvantage is a rather more complex feed geometry over a standard parabolic reflector.

Note that dish diameters of 15 cm or so at millimetric wavelengths compare with diameters of 30–100 cm for microwave dishes.

Link planning

Line of site fixed link design

In point to point radio links, a further consideration at microwave frequencies is to ensure that sufficient ground clearance is maintained across the link.

A transmission path for a UHF/microwave link which is line of sight but only just – i.e. the path close to the ground – can have a refraction loss (also known as loss due to grazing incidence) which exceeds the free space value.

The clearance required to give 'free space' transmission can be calculated by reference to the radio path's Fresnel zones – a zone on a specified surface where the sum of the distance from transmitter to receiver to any point in the zone does not vary by more than half a wavelength.

Figure 12.21 shows a transmitter/receiver path in free space intersected by an imaginary perpendicular plane with concentric circles, A, B, C, representing the path of the secondary waves travelling from the transmitter.

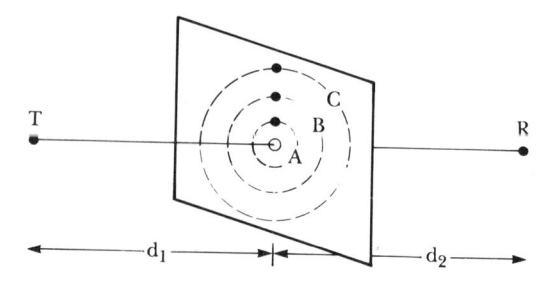

Figure 12.21

Antennas – selection, installation, fault finding and maintenance

The total path length from transmitter to receiver via 'A' is 0.5 of a wavelength longer than the direct path, one wavelength longer via 'B' or 1.5 wavelengths longer via 'C'. The contribution of each successive zone at the receiver is therefore alternately negative and positive.

If the energy of the two zones A and B arrive together at the receiver, the contributions of each of the zones are almost equal and 180° out of phase, producing signal cancellation.

Losses from grazing incidence are typically in the region of 6 dB. The losses can be avoided provided about 55–60 per cent of the first zone around the radio path is kept free from obstruction. Designers of point to point links try to achieve a tower height which keeps close to 100 per cent of the first zone unobstructed, though there is little benefit once the 55–60 per cent has been achieved. In practice, the clearance needs to be a few feet at the ends of the path up to 50 feet or more at mid path.

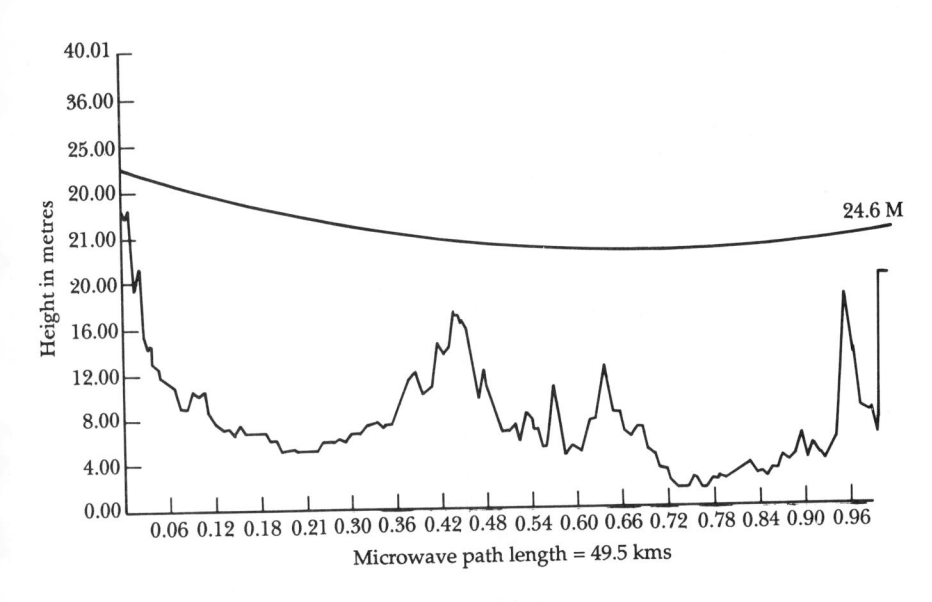

Figure 12.22 *A typical microwave profile path*

The radius of the first Fresnel zone for a specific fequency can be found from:

$$r = \frac{\lambda\, d1 - d2}{d1 + d2}$$

where r = radius (in metres)
  λ = wavelength (in metres)
  d1 d2 = distance as in diagram (in metres)

The calculations are strictly only valid for horizontal polarities, but in

Mobile Radio Servicing Handbook

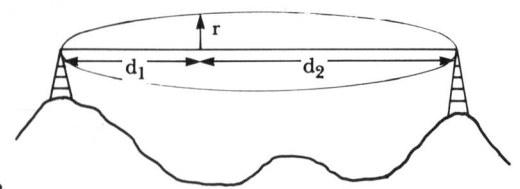

Figure 12.23

practice can be applied to UHF-microwave frequencies with reasonable accuracy.

Earth curvature correction

Path profiles are normally drawn to take account of the standard refraction equivalent to 4/3 rds of effective earth radius using special curved graph paper 4/3 rds effective earth radius is also known as a k factor of 1.33, i.e. a factor which assumes an earth radius 33 per cent larger than actual.

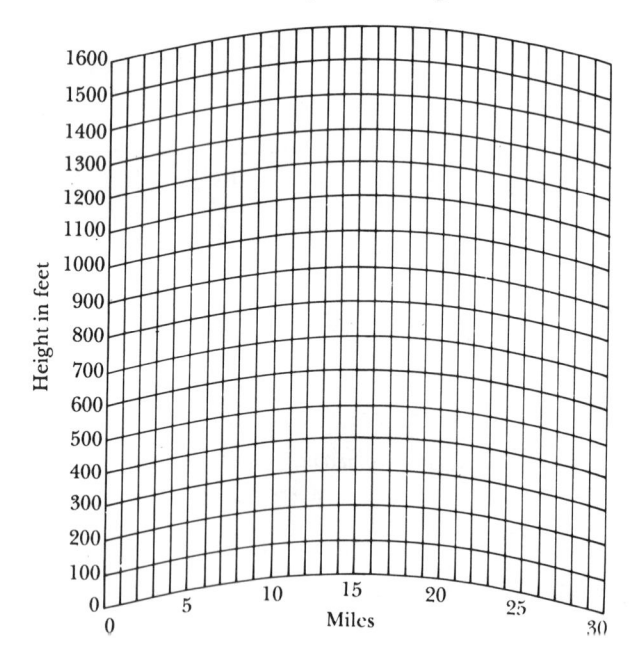

Figure 12.24

It has been suggested that a lower k factor of 0.7 may be more appropriate for planning critical high reliability links. A 0.7 k factor assumes an earth radius 30 per cent smaller than actual, i.e. a steeper curvature requiring additional clearance. This calculation assumes the refractive index causes the wave path to have a curvature *opposite* to that of the surface of the earth.

Figure 12.25 *Typical multi-coupled PMR site on a BBC booster transmitter mast (Photograph Courtesy of Aerial Sites Ltd, Chesham, Bucks) shared with a cellular installation and microwave links.*

13 Typical operator and system problems

Test calls

From both the operator's and system maintenance point of view, it is important to establish a reporting system in which relative deterioration of reception with a given area can be identified.

When a supplier originally installs a system, it should be normal for reception tests to be carried out across the service area, with report forms establishing the base station, mobile, vehicle location and direction and giving an assessment of signal quality.

Specifying a received signal level (say 5 μV PD) at the base of a roof mounted quarter wave antenna across a percentage of the service area can provide a basis for *objective* comparison. *Subjective* assessment can be made on a quality rating (good to bad to unintelligible).

If no original documentation exists, it can still be worthwhile establishing an in-house system with a routine of test calls to monitor coverage and reception standards.

Phenomena of which to be aware (i.e. problems not created by hardware or system faults) include cloud and rain static, in which low clouds discharge into the base station antenna (symptoms are a ticking sound or whine). (This can sometimes be cured by grounding the antenna.)

Common hardware or system faults can include:

- Misaligned receivers resulting in poor sensitivity; causes poor signal to noise ratio.
- Squelch threshholds set too high; will only open when presented with an unrealistically strong signal, often leads to complaints (unjustified) of poor sensitivity.
- Transmitter alignment. Misalignment in the power amplifier chain of a transmitter can reduce power output, as can poor quality connectors (including water ingress) or incorrect antenna installation.
- Transmitter off frequency. Symptoms of drift and spurious signals and reduced power or frequency.

Note that frequency drift can be a problem in quasi-synchronous systems, because of the system's dependence on closely controlled phase relationships.

Crystals most commonly 'age down', i.e. drift exponentially down in frequency through life. Occasionally crystals will 'age up'. A user can specify the direction of drift when purchasing the crystals. It is generally recommended that all crystals for a system should be purchased together, including spares, and be used together and continuously to maintain consistency. Crystal 'modules' are also becoming available in which the ageing characteristics are corrected through life by using software correction techniques.

Batteries

The most common faults tend to be related to batteries within portables, often wrongly attributed to poor base station performance.

Nickel cadmium or ni–cad batteries are the most commonly used type of battery for portable equipment today. Their power to weight ratio is good, the number of cycles over which they can be charged and discharged is usually in the area of five to six hundred, and they have a shelf life, when not in use, of at least two years. The disadvantage of these batteries is that they are fairly temperature sensitive, i.e. when the temperature falls below 0°C problems can occur in the charge and discharge characteristics of the battery. The other disadvantage is that the charger circuit, either in the equipment or separate, needs to be a fairly complex piece of circuitry to control the charge condition correctly.

Lead acid batteries form the other main type in use today. These are sealed lead acid batteries, not the open type of battery as used in cars etc. The lead acid battery is less sensitive to temperature than the ni–cad, and can use an extremely simple charging circuit. One other advantage is that it has a very low self-discharge rate, which means that when sitting on the shelf, not used, in a charged state, it can maintain itself for many months without a re-charge. It does, however, have a shorter shelf life, only some fourteen to sixteen months.

Other types of battery in use include:

1 Alkaline cells
 These are not re-chargeable but have a very high power to weight ratio and a shelf life of some twenty-four months, but are expensive.
2 Mercury cells
 Mainly used for very miniature equipment, such as calculators, watches etc.
3 Lithium cells
 As yet not a wide commercial application, primarily used in the military field. High voltage capability (3 V or more), wide operating temperature range, good power density and flat stable charge profile, and a shelf life of over ten years; disadvantage is primarily cost.
4 Lithium thionyl chloride
 A specialist application variant of lithium cells, used for long term

emergency standby applications, with a twenty-five year inert storage shelf life.

Nickel cadmium batteries can be expected to have an endurance of between 8–10 hours, but this is very dependent on duty cycle. They have the advantage of maintaining their power output but the disadvantage that the output falls rapidly (and suddenly) at the end of the endurance period.

Ni–cad batteries also require careful recharging. Top-up charging from a semi-discharged state can give the battery a memory – next time round it thinks it has discharged when it reaches its previous discharge level. The rule of thumb for charging time is 1.4 times the discharge time; in practice (reference J Davies *Private Mobile Radio – A Practical Handbook*) a recharge cycle of fourteen hours against a discharge period of ten hours has been found to be effective, but still results in an operational life of no longer than eighteen months, with an economic phased replacement cycle of fifteen months.

Mobiles have a particular requirement in that the power needed in transmit is a factor of ten times greater than the power needed to receive.

Batteries, therefore, have to be capable of working at high power densities. Ni–cad batteries typically deliver 30 WH per kg. Nickel cobalt is a possible alternative and offers the potential of delivering 55–60 WH per kg.

As a rule of thumb, 2 W of primary power is needed for every 1 W of transmitted power.

Health and safety

When establishing or re-equipping a workshop, it is important to consider the health and safety aspects of the working environment, taking into account factors such as earthing of equipment, protection and trip circuits, type of benches and material (wood is still often preferred), spacing between benches and floor covering.

The Local Health and Safety Officer will be able to provide help and advice. A contact number for the Health and Safety Executive is listed in the appendix.

Service and support philosophy

Manufacturers of modern synthesizer radios are currently claiming mean time between failure rates in excess of 25 000 hours. For a mobile in use for nine hours every day, this would mean one fault every ten years.

In addition, in all radios there will be a small percentage of early failures – infant mortality. These early failures should become apparent prior to delivery during factory soak testing. For radios developing in-service faults, a service workshop would generally be expected to provide a response time to a service call of not more than eight hours. A target repair time in the field would be thirty minutes or less, implying that work is limited to replacing field exchangeable units – PCBs, power supply modules and mechanical items – a 'first line servicing' definition.

Other repairs would require return to a central workshop – a second line service level, which would include validation of the RF performance parameters. Target repair time of a mobile in a workshop should not exceed one hour per mobile.

14 Spectrum efficiency – audio Selcall, trunking and cellular systems

Strategies for spectrum efficiency

In Chapter 2 we stated that only just over 150 MHz of spectrum is allocated in the UK between 30 and 960 MHz for land mobile radio, i.e. 6000 12.5 kHz channel pairs, or 3000 25 kHz channel pairs, to support an anticipated mobile radio user population of several million. Most market estimates suggest a UK user base increasing from approximately 2 million users (PMR, cellular and paging) to 6 million users by the mid to late 1990s.

We can alleviate the spectrum limitation by:

1 Reducing channel widths – we have seen from earlier chapters that, in practice, it is problematical to reduce channel widths to less than 12.5 kHz.
2 Reducing call duration – an advantage of PMR is that average call duration is 20–30 seconds, as opposed to 2–3 minutes on the cellular networks.
3 Sending speech messages as written messages – a speech message taking 30 seconds can be reduced to 2–3 seconds of transmission time using a 1200 baud data over radio transmission rate.
4 Using trunking technologies – to increase the efficiency with which allocated channels are utilized (giving roughly a 25 per cent gain over a community repeater system).
5 Using cellular technologies – to re-use frequencies within relatively close geographic proximity.
6 Using micro-cellular technologies – utilizing digital voice signal processing at millimetre wavelengths.

On a more prosaic level, *simple audio signalling* can be used to reduce the annoyance factor of a number of users sharing a single transmit and receive frequency.

Audio Selcall (selective calling)

In a simple PMR system, say 10–15 mobiles and a base station, all user equipment would normally be tuned to the receive frequency when not transmitting, making all the channel activity audible. We can, however, differentiate between the mobile receiving and the user hearing the channel content by switching on the audio output of the mobile only when its group page signal (SELCALL) is received.

An organization can use Selcall to divide its mobile network into specific groups, sales force, service or security for example, avoiding unnecessary calls.

Transmission can also be stopped if the mobile channel is already engaged.

As the system locks out other users during a conversation, some form of *time out* is normally introduced to limit call duration. When a base station is operated as an unattended repeater triggered by a mobile user, controlled or selective signalling is essential to prevent unauthorized use of the repeater or its triggering by interference.

The main advantage of audio Selcall as a signalling system is that it uses tones within the normal voice frequency range transmitted by the system and can therefore be added to an existing system if required. Because Selcall is transmitted before, not during, speech, the harmonic purity of the Selcall tones is less important than for continuous tones – for instance sub-audio CTCSS. Because tone duration is generally longer than signal fade, audio signalling can be decoded reliably at very low signal to noise ratios. The disadvantage is that maximum information transfer rate is only 90 bits/second.

In its simplest form, a user identity is encoded as a sequence of tones, each tone is a single audio frequency and corresponds with a number. The number of digits in the identity is usually two or five.

Frequency	f_1	f_2	f_3	f_4	f_5
Code	1	2	3	4	5

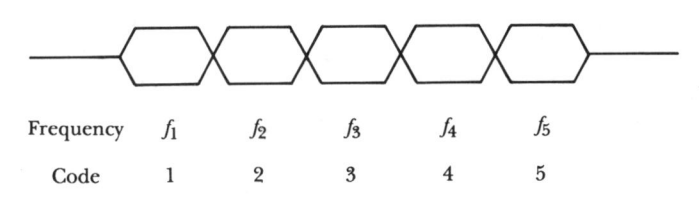

Figure 14.1

If digits are repeated within a sequence, a *repeat tone* is used. It is also commonplace to add a *group tone*, replacing a digit tone and terminating the code. Substituting the group tone in positions 3, 4 and 5 would address a group of 1000, 100 or 10 as required.

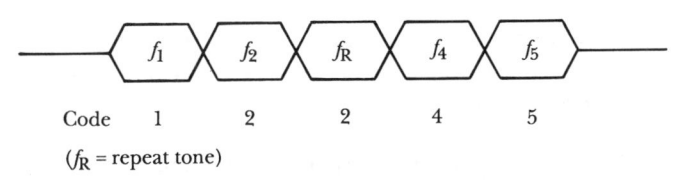

	f_1	f_2	f_R	f_4	f_5
Code	1	2	2	4	5

(f_R = repeat tone)

Figure 14.2

Mobile Radio Servicing Handbook

In large networks involving radio repeaters or pagers which power down to an energy saving standby condition, an *extended first tone* allows time for equipment to power up and stabilize, avoiding any loss of information.

A number of different codes have been established by the various international committees and are as shown in Table 14.1.

Table 14.1 Some typical Selcall codes established by international committees sometimes amended by users or manufacturers.

Standard	ZVEII	EIA	CCIRI	EEA	CCITT	EURO
Tone duration (ms)	70	33	100	40	100	100
Pause duration (ms)	0	0	0	0	0	0
Code 0 (Hz)	2400	600	1981	1981	400	980
Code 1	1060	741	1124	1124	697	903
Code 2	1160	882	1197	1197	770	833
Code 3	1270	1023	1275	1275	852	767
Code 4	1400	1164	1358	1358	941	707
Code 5	1530	1305	1446	1446	1209	652
Code 6	1670	1446	1540	1540	1335	601
Code 7	1830	1587	1640	1640	1477	554
Code 8	2000	1728	1747	1747	1633	511
Code 9	2200	1869	1860	1860	1800	471
Repeat	2600	459	2110	2110	2300	1063
Alarm	2800		2400	2400		
Free tone						1153
Group tone	2400			1055		

ZVEI — Zuverein der Electronisches Industrie. Originally designed to operate on systems with channel separations of 20 kHz, there is often some difficulty in transmitting the highest frequency tone of 2800 Hz on a 12.5 kHz system (the frequency response of the radio being designed to fall off at higher frequencies to reduce band width). This has resulted in a variant known as depressed ZVEI with a tone at 825 Hz.

EIA — Electrical Industries Association

CCIR — Committee Consultatif International Radio Communication. Originally for marine use, requires longer transmission time. Formats with longer timings tend to be more reliable but have a slower call rate.

EEA — Electronic Engineering Association. In frequent use.

CCITT — International Telegraph and Telephone Consultative Committee

See also DTI publication MPT 1316
Other types of signalling used are as follows.

Single burst tone (SBT)

A single audio frequency tone transmitted for a limited period sometimes used in transponding (activating the user's transmitter to send a signal back to the initiator of the call and confirming availability of the radio path).

SBT can be prone to signal imitation, with the receiver responding to an unwanted similar signal caused, for example, by interference.

Dual tone multi-frequency (DTMF) signalling

Eight audio frequencies are assigned to DTMF signalling. To minimize signal imitation, especially by voice, two tones are sent simultaneously and are always selected from the high and low portions of the audio band. The tones are transmitted for a limited period only.

This type of signalling is often used for the transmission of telephone dialling information, used for PABX or PSTN interconnection.

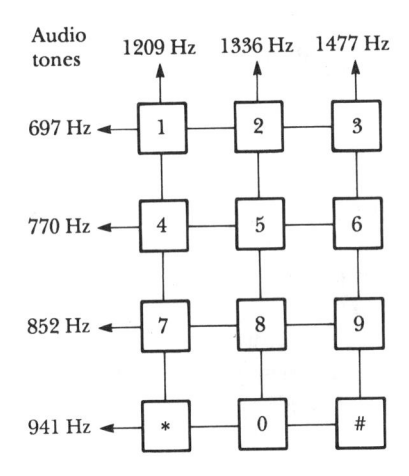

Figure 14.3

Continuous tone controlled signalling system (CTCSS)

A continuous tone controlled signalling system, also known as continuous tone controlled sub-audio squelch.

This system uses nineteen sub-audio tones from a selected series of frequencies below the audio band width of the receiver (below 300 Hz). The tone is operated continuously during speech operation but is not heard by the user.

Such tones are used for the control of receiver squelch and access to repeaters, minimizing the effect of interference. The Code of Practice for CTCSS is contained in DTI publication MPT 1306.

Digital signalling using FFSK

The slow data transfer rate (90 bits per second) sets the application limitations of audio tone signalling. The alternative signalling standard set out in MPT 1317, using fast frequency shift keying, is used to provide us with the basis for data over radio transmissions with a data transfer rate of 1200 bits per second.

In addition to data over radio transmissions, for example data messaging or paging, this faster data transfer data transfer rate allows for the implementation of the relatively complex signalling access and controls required for regional and national trunked radio networks, giving call set-up times of 20-25 seconds or less.

Signalling therefore falls broadly into two categories — in-band and out-of-band. In-band signalling is defined as falling within the standard 300 to 3400 Hz audio band width of the system, and out-of-band either below or above this. Occasionally, the signalling tones are around say 2.8 or 3 kHz, in which case the signalling is out-of-band, and the audio band width is restricted to say 2.5 kHz at the high end. This subsection will attempt to explain how the most widely used signalling tones are generated, modulated and detected.

CTCSS

This is also known as sub-audible tone since the signalling consists of a sinewave tone at between 67 and 250.3 Hz. The DTI have decided upon two groups of sixteen tones within these limits called the A group and the B group. The two groups interleave, and it is illegal to have one system with mixed A and B group tones. The sub-audible tone is sent continually along with each transmission, and is modulated at a low level, around 10 per cent AM or 350 Hz peak deviation on FM. It is picked up at the receiving end by an active filter; another filter removes it from the audio path such that the tone does not reach the loudspeaker.

The tone can be used for a number of control functions, but normally it is used to either unmute a receiver or to key a base station transmitter. The DTI insist on a positive coherent decoded signal to have control of the transmitter key line at a repeater. As such the CTCSS decoder provides this.

One very popular use of CTCSS is on a community base station which is equipped to accept the entire group A or B and re-transmit the same tone as received. In this way, sixteen customer fleets are accommodated on a single base station. Each member of the same fleet has the same CTCSS tone and whenever one of them speaks, the whole fleet can hear and respond. During this time the other fleets associated with the base station are locked out because they are receiving the wrong CTCSS tone. This mechanism is called tonelock, and works by inhibiting the mobile transmitters.

The tone is generated by an active filter with a positive feedback path round it, and in this way the same filter can be used in a transceiver during transmit and receive. In receive mode it is a plain filter with no feedback; in transmit mode the feedback path is added and it oscillates on the same frequency. CTCSS modules are available to retrofit onto virtually any mobile radio, and many are crystal controlled.

Selcall

Selcall is an in-band string of tones sent once at the beginning of a call, to address one mobile or a group of mobiles. Selcall protocol is discussed elsewhere in the chapter. It is generated usually now by a microprocessor in a software loop to provide the correct frequency, and then filtered to make the wave somewhat sinusoidal. It is not so important to keep the distortion down as with CTCSS because it does not accompany any speech transmission and harmonics are not a problem. A typical modulation depth is 60 per cent AM or 60 per cent of system deviation on FM, say 1.5 kHz peak.

Older Selcall systems used an LC oscillator to generate the tones and the inductor had about twelve taps on it to obtain the various frequencies. Detection of the tones is usually done with a custom chip available for the purpose; one popular manufacturer is Consumer Microcircuits Ltd. They are detected by a series of commutating filters which rely on an accurate clock; this is either the microprocessor clock with which the signalling is associated, or an off-chip clock with a single chip decoder. Upon reception of the correct Selcall string, the mobile usually responds by sending it back (called transponding), sometimes with the last 1 or 2 digits modified as an indication of the status of that mobile. The mute will open on the mobile receiver and the transmitter will be enabled, both for a present time, usually reset each time the pressel (press to talk button) is pushed. When receiving a group call, the mobile does not transpond, because if every mobile in the group transponded the base station would be confused to say the least!

FFSK

Fast frequency shift keying is a modulation method to get binary information onto a carrier. As the data is NRZ (non-return to zero), then some way has to be found of representing 0s and 1s by in-band audio tones. With FFSK, a 0 is represented by 1.5 cycles of 1800 Hz, and a 1 by 1.0 cycles of 1200 Hz. The modulation rate is 1200 bits per second, which is equal to the lower of the two tone frequencies and therefore the points at which the data changes state are where the waveform is going through zero. The system is thus phase continuous and as such occupies the least possible band width for the modulation rate, hence the title 'fast'. It is sometimes known as minimum frequency shift keying (MFSK) for the same reason.

It is generated and received by duplex modem chips dedicated to the task; one popular manufacturer is Consumer Microcircuits Ltd, their FX419 chip is very common.

The modulation depth is 1.5 kHz peak, and the method is only used on FM systems.

Line signalling

Line signalling is a huge topic on its own, but here we will just look at a typical AC signalling system. One common requirement is to remotely control a base station over a 600 ohm private wire; usually to key the transmitter and switch talkthrough on and off.

Since British Telecom lines have only 300 to 3400 Hz band width, an AC signalling method is necessary with in-band tones. Older systems used a continuous subcarrier at say 2800 Hz, and curtailed the speech response to 2.5 kHz. This subcarrier was frequency modulated by low frequency tones such as 60 Hz or 80 Hz, and a positive coherent decode of these performed the desired control function at the far end. Later systems send a data stream down the line when required, and an occasional data handshake takes place to make sure the line is OK. A tone may also be sent at a standard level such that the sensitivity of the line termination unit at the far end can adjust itself to changes in line loss. Both systems detect when the line goes faulty; the first by loss of subcarrier, the second by an unsuccessful handshake, and usually the base station will automatically go onto talkthrough under this condition. This is known as 'line fail talkthrough' and provides at least emergency communication.

The subcarrier level in older systems is typically 20 dB below peak speech level on the line, and it is generated by an oscillator whose frequency can be adjusted with a DC voltage (a VCO). The low frequency modulating tones are put onto the VCO control line at a level to obtain about 100 Hz peak deviation. The subcarrier is detected by an active filter at the remote end, and a phaselock loop detector may be attached to this, whose error voltage carries the low frequency modulating tones. Further active filters are connected to the phaselock loop output, one looking for each low frequency tone. The new systems can use FFSK signalling down the line and will choose data transmission times so as to avoid clashes with speech. There is no 2.6 kHz restriction on bandwidth with later systems and clearer speech is the result.

Trunking

Taking the FFSK signalling format as our starting point, the UK trunking

Spectrum efficiency

networks are established on the basis of two national standards. MPT 1327 sets out the required signalling protocols of the system. MPT 1343 sets out the required design parameters of the mobiles. The eventual intended advantage of this standardization is that a user will be able to transfer (and possibly roam) freely between the networks, i.e. the standards will be used as an air-interface specification.

The standards are not exclusively for use on Band 3 frequencies and can be applied in principle to any land mobile trunking system, but for the sake of explanation, we will take the UK national Band 3 networks as an illustrative example.

We have said that a trunking network typically provides 25 per cent more capacity than a conventional community repeater system, one extra voice channel every five channels, four extra voice channels in a twenty channel system. In addition, within a wide area or national network, significant frequency re-use becomes possible.

System efficiency is achieved by a process of queueing and dynamic channel allocation and control managed by a central system controller – the human analogy is that of the queue in your local post office or bank – with one disciplined queue for several counters (generic queueing).

In the trunked radio network, the aim is to ensure that the calling party and the called party reach the top of the respective queue (or stack) at the same time. When, and only when, both parties are ready for the call, the system will send a 'go to channel' instruction, allocating an available traffic channel for the call.

We can see how this can be implemented on a national scale. Figure 14.4 sets out the relatively simple hierarchy of the first 'regional' stage of one of the two UK Band 3 national networks. The hierarchy is fundamentally that of a community repeater, with radio despatcher access from the client's radio base

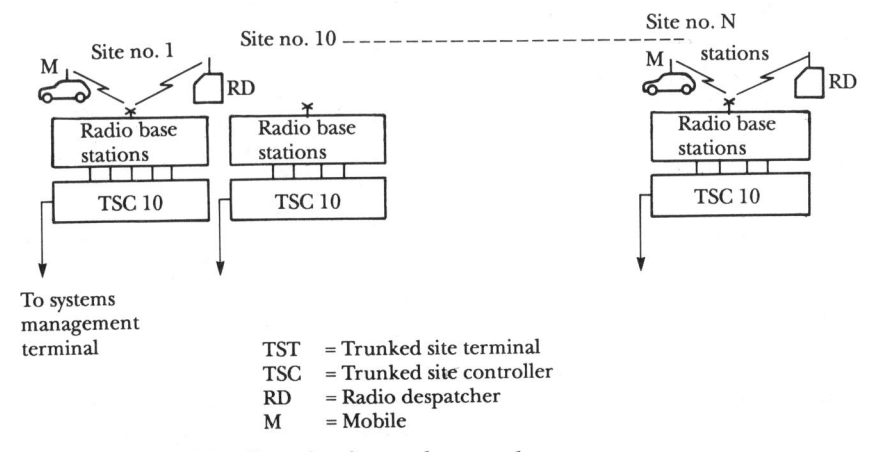

TST = Trunked site terminal
TSC = Trunked site controller
RD = Radio despatcher
M   = Mobile

Figure 14.4 *Band 3 radio national network – stage 1*

Mobile Radio Servicing Handbook

to the central radio base. The digital signalling protocol allows for additional facilities to be offered – assignment of a dedicated channel, call queueing, call status messages, short form dialling and an indication of the source of the incoming call. Cover is, by definition, that provided by the single central radio base.

The hierarchy in stage 2 provides for interconnection between regions, with a node controller per region, the sharing of channels across the regions and the availability of a simple line despatcher link from the client's base. In addition, the network allows an overlaying of a 184 bit data message on the control channel used for messaging/overlay paging.

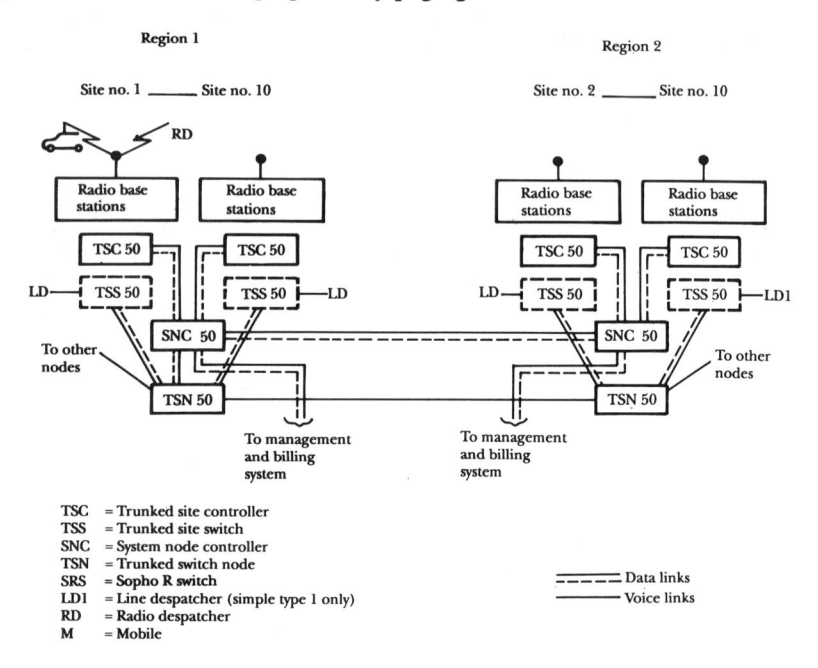

Figure 14.5 *Band 3 radio national network – stage 2*

Stage 3 is a fully fledged national network with sharing of channels (in this case 100 channels) on a national basis, access to the client's PABX and limited (short form dialling) access to the PSTN.

On a national basis, a frequency re-use factor of 5.56:1 is said to be achievable, i.e. 556 channels of which 72 channels need to be used for control purposes, giving the capability to support a system capacity of up to 60 000 users.

This type of trunking technology shares a number of common features with the cellular networks. The central controller sends out a regular 'invitation to make a request' signal to all the users within the system. Each radio has a unique electronic serial number which the controller identifies, together with

Spectrum efficiency

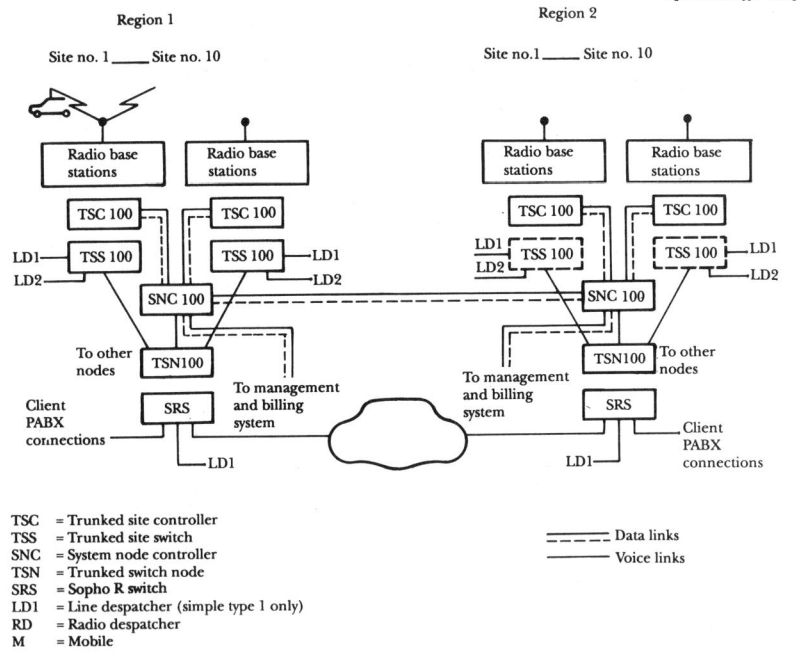

Figure 14.6 *Band 3 radio national network – stage 3*

the location, the validity (has the user paid his monthly subscription and/or air time bill) and status; the user will either be switched off, on stand-by or ready to make a call.

In a typical cellular network such as UK TACS, the system is fully duplex. Polling of the mobiles takes place on a permanently allocated forward control channel (FOCC), which provides the opportunity for the mobile to make a call. Access availability is indicated by a continuous busy/idle bit stream.

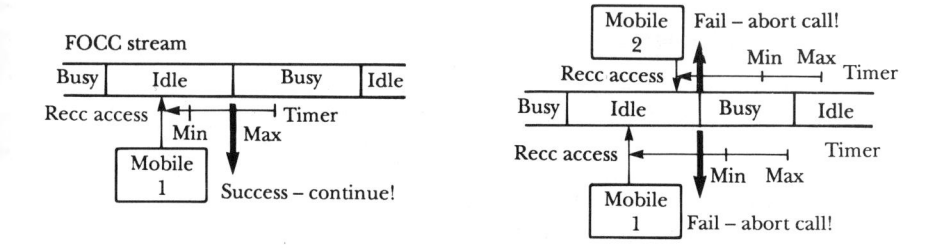

Figure 14.7

Within a semi-duplex trunking network, the system delegates a channel for use as a control channel and will poll (ALOHA) the mobiles, indicating a number of defined forward time slots available for access, the number of forward time slots reduces as the system becomes busier.

If the user mobile wishes to make a call, it will send a request (REQ) with its mobile number and identity and required call destination (another mobile, PABX or PSTN number).

The system responds by repeating the information (REP) setting the next time slot on the receive channel and requesting the called mobile to acknowledge (ACK) on the transmit channel. If the called unit is available and there is a spare traffic channel, the system sends a 'go to channel' instruction to both parties to transfer to a traffic channel. If no traffic channel is available, the system will set up an automatic re-try sequence or cleardown instruction.

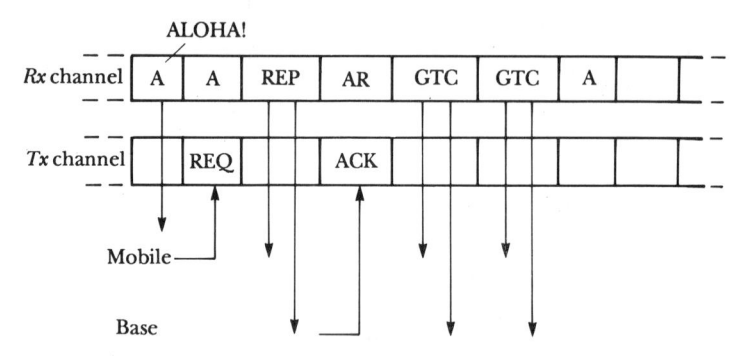

```
                        ALOHA!
 Rx channel | A | A | REP | AR | GTC | GTC | A |

 Tx channel |   | REQ |    | ACK |    |    |   |

              Mobile

              Base
```

Figure 14.8

The whole 'handshake' routine is shown in Figure 14.9 together with bit timing (at the 1200 bit/sec baud rate 6 bits equals 5 milliseconds).

LET (link establishment time) allows the mobile to power up, PRE (preamble) and SYNC (synchronization) ensures correct time slot alignment. ADD 1 provides an error parity check and CCSC (control channel system

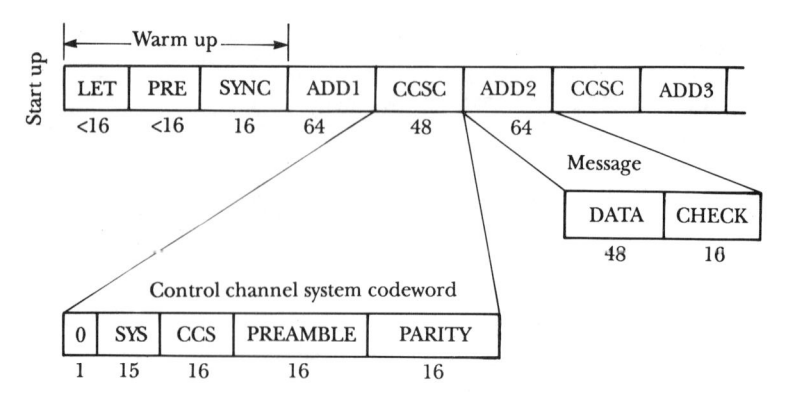

```
          |— Warm up —|
Start up | LET | PRE | SYNC | ADD1 | CCSC | ADD2 | CCSC | ADD3 |
           <16   <16    16     64     48     64

                                            Message
                                          | DATA | CHECK |
                                             48     16

         Control channel system codeword
         | 0 | SYS | CCS | PREAMBLE | PARITY |
           1    15    16       16        16
```

Figure 14.9

Spectrum efficiency

codewords) contain the transmitter and system identity codes and system code (SYS) with an additional parity check. (For a more detailed explanation of parity checking see chapter 16 Data over radio). ADD 2 continues with a 48 bit stream of data and 16 bit parity check and so on.

Signalling can also be carried out during a call by using blank and burst signalling (short bursts of data interleaved with speech).

The central system controller is linked to a data base containing all the status information on users within the system. The data base also holds information on the performance parameters of the network in terms of failed called analysis, 'busy hour' channel loading and service levels.

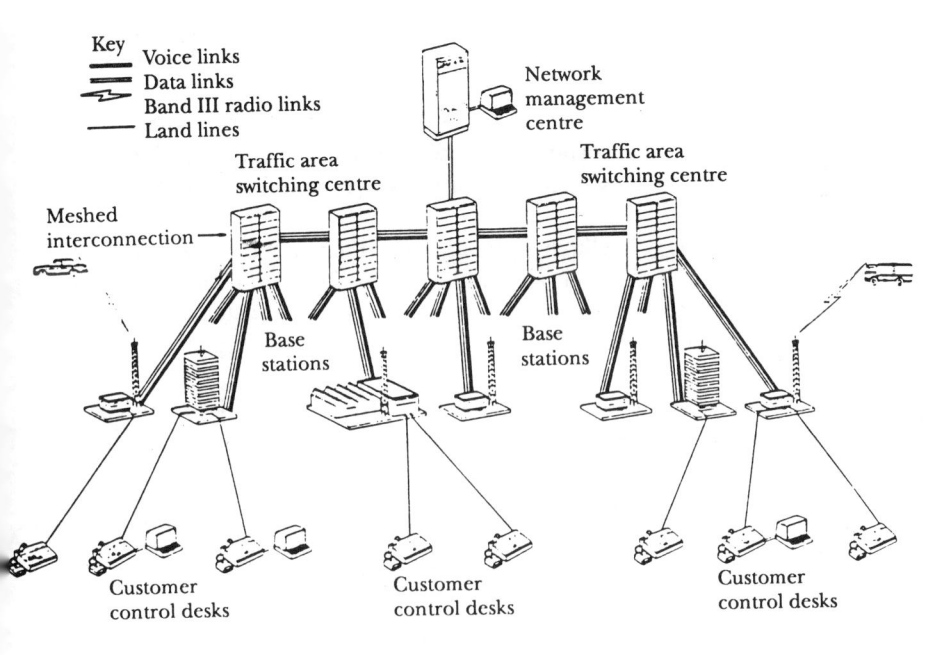

Figure 14.10 *GEC National One-Band 3 network hierarchy providing national UK coverage from ninety base stations*

Cellular radio

Cellular networks add an additional hierarchy of complexity to a trunking network. The philosophy still depends on a central system controller knowing where the mobiles are and their status, but with a number of additional control parameters:

1 Cellular systems provide automatic hand off from cell to cell. The system therefore has to be able to monitor signal strength and to be able to instruct mobiles to change channel at the cell boundary.

2 To maximize frequency re-use within the cellular honeycomb, a mobile will be instructed to transmit at the lowest possible power level necessary to achieve an acceptable communication quality.
3 Because users are billed for air time and because PSTN interconnection is an integral part of the network, the system has to capture and process the necessary billing information for each call.

In the UK TACS structure, the complexity of the signalling protocol is such that the 1200 baud rate signalling used in trunking systems would result in unacceptably long call set-up times. Signalling is therefore carried out at a rate of 8000 bits per second.

FFSK becomes impractical at this signalling speed, it becomes difficult to design circuits to encode and decode strings of 1s and 0s, and difficult to apply FFSK as a modulation technique.

The solution is to Manchester encode the data (ones are represented as 1/0, zeros are represented as 0/1) and the informaton is modulated on to the carrier using phase shift keying (see chapter 16 Data over radio) for a definition of PSK).

This is not a universal cellular signalling solution and we can see that both signalling and parameters such as modulation method and channel spacing, vary significantly across the current European analogue systems. Note that higher signalling speeds generally also demand wider channel spacing of 20, 25 or 30 kHz.

The following chart gives an indication of the incompatibility between the current FM analogue systems.

System specification	NMT*	AMPS	TACS	Japan NTT	Radiocom 2000	Network C
Number of RF channels	180 or 220	666	1000	600	256	22
Channel spacing kHz	25	30	25	25	12.5	20
Duplex spacing MHz	10	45	45	55	10	10
Signalling speed (bit/sec)	1200	10000	8000	300	1200	5280
Subcarrier modulation	FFSK	PSK	PSK	PSK	FFSK	Direct carrier keying NRZ
Frequency range (MHz)	454–468	825–890	890–960	870–940	406–430	451–466

Spectrum efficiency

Countries adopting, or about to adopt, the various standards are as follows:

NMT*	AMPS	TACS	Japan NTT	Radiocom 2000	Network C
Scandi-navia	USA	UK	Japan	France	Germany
Austria	Canada	Hong Kong			South Africa
Netherlands	Australia	Malta			
Switzerland	Sri Lanka	Italy			
Saudi Arabia	Brunei	Spain			
Malaysia	Thailand				
Belgium					
Luxembourg					
Tunisia					

* Now also NMT 900 (Nordic Mobile Telephone)
AMPS advanced mobile phone service
TACS total access communications system

The systems do, however, share a familiar format of signalling protocol known as the *data telegram*.

We have established that each base station

1 needs to know which mobiles are in its area;
2 needs to be able to monitor communication quality; and
3 must be able to process information on billing/access clearance.

A typical data telegram containing
  N1 – power level
    N2 N3 – channel number
      P – traffic channel allocation
        YZX – PSTN codes/subscriber numbers
          NABC – new data

Bit synchronization	Frame Synchronization	N1	N2	N3	P	Y1	Y2	Z	x1	x2	x3	x4	x5	x6	Na	Nb	Nc
101010101010101	11100010010	6	1	3	5	5	3	5	7	6	1	6	5	3	2	1	3
101010101010101	11100010010	0110	0001	0011	0101	0101	0011	0101	0111	0110	0001	0110	0101	0011	0010	0001	0011

Figure 14.11 *Typical data telegram*

The mobiles are periodically requested to transmit a 'data telegram', a bit stream of data containing the call number, serial number and maximum RF power of the mobile.

The base station can instruct the mobile to change RF channel and/or power level, typically from 10 watts down to a few milliwatts.

In addition within the AMPS and UK TACS cellular structure, there are supervisory audio tones (SATS) which are transmitted currently with voice transmission. Three frequencies are used:

5970 Hz
6000 Hz
6030 Hz

to distinguish between different base stations transmitting on the same RF channel.

A signalling tone (ST) at 8 kHz is also added to perform a function similar to on/off hook signalling in a conventional telephone network, a tone lasting 1.8 seconds to indicate termination of the call.

UK TACS structure – spectrum utilization

The two UK network operators – Racal Vodafone and Cellnet – between them, support the largest cellular mobile radio user-base in Europe.

Taking the Racal network in 1989 as an example of spectrum utilisation potential, an allocation of 300 channels (with 150 extra channels known as E-TACS in the London area) is shared across 362 base stations to give 659 cells and 7706 voice channels to support 250 000 users.

Cell sectorization – using directional antenna to sub-divide cells – is being applied to help support the demand for additional air time. Racal claim that their network could, theoretically, ultimately support up to one million users with adequate service levels.

Implications for testing and in-service support

A mobile operating incorrectly within a cellular network would create chaos. A 'maverick' mobile will automatically be banned from the system. As soon as a radio begins to operate outside its defined parameters, it will need to be repaired.

Typical repair routines include the following:
1 Signalling tests
2 RF parameter tests
3 Audio tests
4 DC checks

Signalling tests

These can be divided into first line testing, the type of test carried out on a mobile in service to establish whether the unit is operating within defined signalling parameters, and second line testing which is the type of test carried out to validate signalling protocols at time of commissioning and for system diagnostic work.

First line testing of signalling This involves the simulation of a land station communicating with the unit under test (UUT) via a cable connected to the antenna socket, transmitting and receiving any command sequence, supervisory audio tone (SAT) or signalling tone (ST) which a land station would normally produce, within the following parameters:

Transmitted output level	A minimum of 1 mV into 50 ohms (−47 dBm)
Input sensitivity (RF)	A minimum of 0.1 mV (−127 dBM)
Receiver frequency range	890.0 MHz to 915.0 MHz
Transmit frequency range	935.0 MHz to 960.0 MHz
Channel spacing	25 kHz (every channel)
Modulation/demodulation	8 K bit BCH Manchester FSK encoded data +3 SAT and 1 ST (FM)
Receiver audio band width	300 Hz to 3.4 kHz
Supervisory audio tones	5.97, 6.00 and 6.03 kHz ±1 Hz
SAT deviation	±1.7 kHz
ST deviation (8 kHz)	±6.4 kHz
8 K bit FSK deviation	±6.4 kHz

A typical test routine would be as follows:

```
Power on self test
DISPLAY "SELECT TEST REQUIRED
   TEST A
   TEST B
   TEST C"
TEST C selected (auto test)
Send control channel overhead message requesting mobile to read and
display: Mobile ID, serial number and called number
Page mobile
Voice channel audit
Measure SATs
Perform hand-off channel 1 to 300 to 600 to 800 to 1000
Alert mobile
Direct operator to answer mobile, and listen for audio test pattern
Flash test (dial a number)
Replace handset
```

Manual release detected
END OF AUTOMATIC TEST

First line testing would not normally include RF and audio tests.

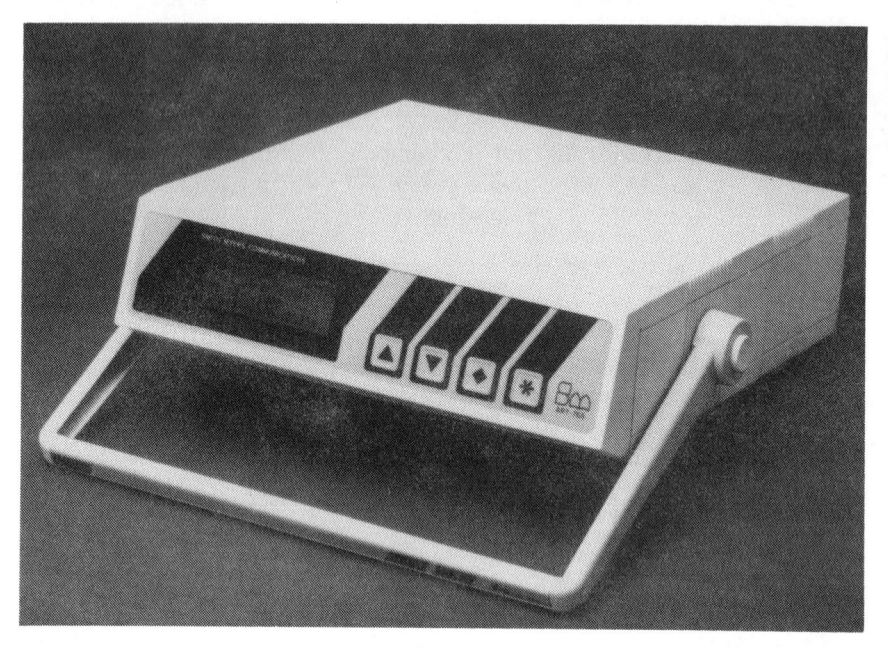

Figure 14.12 *Typical first line test unit used to check signalling protocol on a TACS mobile*

Second line testing of signalling At a second line test level, it would be usual to include error generation capability to test the error correction protocols within the system (the test module increases the number of bit errors artificially implanted until the mobile refuses the call).

RF parameter tests – second line testing

Critical RF parameters which need to be measured include:

Frequency The ability to measure the frequency of the supervisory audio tones and the signalling tone. The ability to measure carrier frequency.

Modulation – modulation depth/deviation The ability to measure the modulation depth/deviation of the supervisory audio tones and the signalling tone. The ability to measure the modulation depth/deviation of the audio signal. The ability to measure the modulation depth/deviation of the data telegram.

With the UK TACS cellular structure the deviation measurement of the data telegram has to be made within a fifty millisecond time scale requiring a frequency pre-set (the modulation meter has to be pre-set to the channel frequency on which the cellular mobile will reply), and a deviation hold facility within the test equipment.

RF power The ability to generate RF at varying levels to establish the call threshold performance of the system, plus the ability to be able to repeat call origination at the sensitivity limit of the mobile to establish the probability of a successful call as a quality criterion.

Audio tests

One of the most fundamental measurements on the receiver is to establish maximum usable sensitivity (MUS), specified as the minimum level of a modulated carrier signal applied to the receiver antenna terminals which will produce an audio signal of a certain minimum quality. The required quality is defined in terms of a 20 dB SINAD measured through a noise weighting filter as specified in CCITT P53A.

DC checks

An ability to measure DC, typically from 300 mV to 30 V and AC from 10 mV to 30 V.

Other points on test equipment which need to be taken into account include:

Spectral purity

Important when selectivity tests are being made on two signal measurements, where the interfering signal outside the receive channel must be spectrally pure.

Modulation measurement

Modulation meters need to have a demodulation bandwidth capable of handling audio and digital modulation.

Summary

It can be seen that the RF operational parameters of any cellular system require careful control – ultimately any mobile communication system based on analogue transmission techniques is limited by the voltage, temperature and noise constraints of the analogue signal environment, hence the need to look at a longer term digital alternative.

15 Pan-European digital technology

Digital speech and digital modulation/processing techniques

The digital techniques proposed for the new Pan-European digital mobile networks exploit the redundancy inherent in the human voice; provided the sampling rate is fast enough, the delay mechanisms of the human ear restore a reasonable representation of the original unsampled message, in the same way that cine or video film utilizes the delay characteristics of the human eye.

Digital modulation involves a combination of techniques which have to work within the following constraints.

Transmission data rate

This should not exceed 16 K bits per second for the system to be reasonably spectrum efficient.

Delay

Delay should be less than 8 msec.

Non-speech capability

The demodulation process has to be capable of separating out data bit streams from voice bit streams.

Speech synthesis

Much of the intelligence/information in speech is carried by *low frequency modulation* of individual carrier frequencies. This provides the basis for speech synthesis techniques.

Dudley's Vocoder of 1936 provides an early example. Dudley divided speech into ten channels – each with a 300 Hz bandwidth. The low frequency modulations in each band were extracted by envelope modulation using diodes followed by 25 Hz low pass filters. Within each channel, the system was designed to detect a hiss or a buzz. At the receiver, a noise generator was switched to reproduce a hiss or a buzz at the approximate frequency and as a square wave. The result was intelligible speech with a 10 times reduction in bandwidth (from 3 kHz to under 300 Hz).

In modern speech synthesis, filters are used to determine the frequency and amplitude of vowel sounds (AEIOU), consonants, nasal sounds (M and N), and fricatives (S, Z and TH). These speech waveform characteristics are then expressed in terms of filter co-efficients.

In Pan-European (GSM) digital cellular networks, the filter co-efficients are extracted from the input speech signal, and transmitted at 3.6 kbps, accompanied by an 'excitation' sequence at 9.4 kbps (a data transmission rate of 13 kbps).

At the receiver, the filter co-efficient and excitation sequence are re-assembled to reconstruct an (accurate) representation of the harmonic structure and spectral envelope of the original input speech signal.

Speech codecs have been developed for military applications with data transmission rates as low as 2.4 kbps.

Note that codecs do not necessarily work equally well in all languages and may have to be practically tested and optimised for the specific countries of application.

Signal preparation

Whatever process is used to encode and decode the speech waveform, there remains the additional problem in radio of overcoming the shortcomings of the radio link. As a general principle, as data transmit speed increases, the bit error rate rises. As the data rate increases, or the frequency decreases the effect of ignition noise increases. We need to find a method of error correction which ensures that sufficient 'intelligence' will arrive at the receiver to provide the basis for reconstruction of the signal. This is achieved by:

1 Channel coding. Channel coding reduces the effect of fading, and reduces the need, for example, for diversity schemes. Channel coding involves the addition of 'redundant' code to the signal, e.g. a 15 kHz signal would be accompanied by 10 kHz of redundant code within 25 kHz.
2 Interleaving. Interleaving is a technique which helps to disperse errors that occur in clusters. Bits from several code words are interleaved prior to transmission.

Speech quality

One problem with digital radio systems is to match the dynamic range of existing systems. The analogue PSTN provides a dynamic range of between 40 and 70 dB, the digital PSTN provides 25 dB. UK TACS with an FM deviation of 9.5 kHz provides 30 dB. Speech quality improvements can be achieved by companding – a technique whereby the speech signal is compressed prior to transmission, and expanded in the receiver – a ratio of 2:1 is achievable with today's companders while retaining adequate linearity.

A delay of only 5ms is regarded as acceptable within analogue TACS at the point of PSTN interconneciton.

Existing digital speech-processing methods build in considerably longer delays, for example a speech codec can take 10ms, channel coding can take between 20 and 50ms multiple access techniques add another 10ms. A delay of 200 ms would cause speakers to interrupt one another.

Modulation techniques

The modulation technique has to maximize bandwidth efficiency, the ratio of data rate to channel bandwidth in Gps/Hz and avoid an abrupt waveform transition at the moment of data change, to avoid spreading of the spectrum baseband.

Modulation alternatives include:

QPSK	Quaternary phase shift keying
16Q–AM	16 and 64 level quadrature amplitude modulation
64Q–AM	
MSK	Minimum shift keying
TFM	Tamed frequency modulation
GM	Gaussian modulation
CPSK	Correlative phase shift keying

The Pan-European (GSM) network will use Gaussian Minimum shift keying with a 0.3 modulation index. A more detailed explanation of modulation techniques is given in Chapter 16.

Transmission techniques

Transmission techniques include:

FDMA	Frequency division multiple access (similar to analogue FM)
TDMA	Time division multiple access. TDMA separates receive and transmit bursts into short time slots. The advantages of TDMA are the elimination of the duplex filter in the receiver, and generally less critical filter selectivity, frequency stability and VCO noise parameters.

Wide band v narrow band

Narrow band TDMA (200 kHz) is regarded as the most satisfactory solution for the Pan-European network, with a lower transmit power requirement, and less complex signalling processing compared with wide band (several MHz).

The implementation of matched filters for a large number of spread spectrum codes is an additional problem within wide band, as would be the legislative allocation of such broad segments of spectrum.

The Pan-European GSM (group special mobile) network is to have band width and channel spacing of 200 kHz, and will use digital voice coding at 13 kbps (kilobits per second) per channel. The time division multiple access will use time slots of less than 1 msec to provide eight channels per carrier for two frequency duplex.

Reception/demodulation techniques

Decoders are becoming increasingly intelligent. *Soft decision decoders* in particular improve detection/demodulation accuracy by deciding whether a signal is above or below the decision threshold and computing a 'confidence' number (representing nearness to, or distance from, the threshold) producing a 'weighted average' demodulation.

Future techniques

Techniques for improving radio reception in general include combining two or more versions of the transmitted signal at the receiver, space, time, field, polarization or sideband diversity comparisons, producing a 'best fit' signal for demodulation/reconstruction.

Signalling control protocols

The Pan-European (GSM) network will have additional levels of complexity over and above the national analogue cellular networks, created by the need to keep track of mobiles crossing national frontiers (known as visitor location), and to manage trans-national billing of air time.

The signalling protocols are being established by the GSM (Group Special Mobile) committee reporting to CEPT (Conference of European Posts and Telecommunications) – the signalling system is known as CCITT signalling System Number 7.

Channel loading implications

Traffic density is the prime mover behind the drive to introduce a digital network. By the mid 1990s there will be some twenty-two million cars in the UK with 25 per cent of them concentrated within the 2000 square kilometres of the M25 circle.

Digital transmission techniques, both FDMA and TDMA, offer better immunity to interference, allowing the same frequency to be re-used more often in the cell plan, doubling the traffic density of a comparable analogue system for this reason alone.

We can compare the traffic carrying capability of the systems in terms of Erlangs – a unit of measurement named after the Danish scientist Erlang (published work on telephone traffic theory between 1909 and 1920).

$$\text{Traffic density} = \frac{MCT}{3600}$$

in which M = number of mobiles, C = call rate (calls per mobile per hour), T = average effective message duration in seconds.

1 Erlang, for example = 100 per cent density
0.3 Erlangs = 30 per cent density

Density figures exceeding 40 per cent in a simple system will result in a noticeable deterioration in service level, so the number of potential users supportable has to be calculated on this basis.

The analogue TACS 12 cell repeat pattern provides a loading capability of 15 Erlangs per square kilometre, equivalent to 3.3 channels per cell per MHz. In practice, the existing cellular system can support up to 1500–1700 users per square kilometre with acceptable service levels/access times.

Time division multiple access schemes are regarded as being practical for use in cell sizes down to a radius of 100 metres (micro-cellular), and would give a channel loading capability of up to 200 Erlangs per square kilometre. This would provide the necessary infrastructure for a six million user base (10 per cent of the UK population) by the mid 1990s.

CT2 phone zone cordless telephone applications

TDMA is also to be applied to the new CT2 cordless telephone standard. Speech is digitized at 32 kbps (kilobits per second) and transmitted at 64 kbps. Each 2 MS of speech is transmitted in 1 MS with the 'gap' used for the speech return path. The system has the advantage of avoiding the need for 'paired frequencies' (typically 45 MHz apart) and the need for an expensive duplex filter (see table 15.1).

Pan-European digital technology

Table 15.1

Summary of digital speech – digital modulation techniques					
Digitization of speech→	Signal preparation→		Modulation→	Transmission techniques→	Reception/demodulation
Wave form coding	Synthesis	Channel coding	Quaternary phase Shift keying	Frequency division multiple access FDMA	Echo cancelling
Pulse code modulation	Sub-band coding	Interleaving	16 and 64 level quadrature amplitude modulation	Time division multiple access TDMA	Soft decision decoding
Delta modulation	Linear path prediction	Companding	MSK minimum phase shift keying		
64K Bits	16K Bits				

Summary

Speech synthesis and digital modulation techniques in general are made possible by the 'enabling' technology of VLSI (very large scale integration) and ULSI (ultra large scale integration) providing the capability to process large amounts of digital information at high speed with minimal power consumption.

Digital modulation and processing techniques will deliver significant benefits in terms of channel loading, and allow for the integration of voice and data services with the PSTN integrated services digital network, ISDN.

The future

1991 has been heralded as the start date for the Pan-European (GSM) mobile radio system. CT2 phone zones/telepoints are expected to be introduced in parallel.

Spectrum for GSM has been reserved between 935 and 960 MHz and 890 to 915 MHz.

The system will have the capability of handling voice and non-voice services, requiring different bit rates.

Each individual channel will be allocated a rate of 64 kbps. The GSM speech codec working at 13 kbps will therefore allow up to four voice conversations per channel.

Mobiles can be instructed to frequency hop up to 217 times a second, randomising the effects of interference.

In the longer term, digital signal processing will provide the basis for implementing cost effective two-way radio at the higher (millimetric) frequencies, providing a micro-cellular mass user system structure for the early years of the twenty-first century – work in this area is being co-ordinated by a

European joint development programme – RACE (Research and Development in Advanced Communications in Europe). In the shorter term we are likely to see broad band mobile communication networks introduced between 1.7–2.3 GHz.

16 Data over radio

Introduction

Data over radio – the sending of written text over a radio link – is already playing a major role in our analogue mobile radio networks and has a major advantage in terms of air time utilization – a speech message of 30 seconds can be sent as a written message in 3 seconds.

This Chapter sets out the techniques, terminology and technology of data messaging over radio.

Error detection/correction techniques and the choice of shift keyed modulation methods, are equally applicable to the transmission of information from a speech codec.

There are, therefore, a number of references between this chapter and the preceding Chapter 15 on Pan-European digital technology.

ASCII 'strings'

ASCII (American Standard for Information Interchange) is the common alpha numeric system applied in messaging and data over radio applications. The system provides 128 possibilities using a 7-bit code.

Examples:

Decimal		Binary 7-bit code						
A	65	0	1	0	0	0	0	0	1
B	66	0	1	0	0	0	0	1	0
C	67	0	1	0	0	0	0	1	1
D	68	0	1	0	0	0	1	0	0
E	69	0	1	0	0	0	1	0	1
			64	32	16	8	4	2	1

		Parity check						
0	48	0	0	1	1	0	0	0	0
1	49	0	0	1	1	0	0	0	1
2	50	0	0	1	1	0	0	1	0
3	51	0	0	1	1	0	0	1	1

Mobile Radio Servicing Handbook
Parity checks

Digital information sent over a mobile radio can be subject to errors (caused by Rayleigh fading and interference) and an error detection method is required.

This is achieved by adding *redundant* bits either to the beginning or end of the data information block.

The simplest form of redundancy is the parity check in which an extra bit is added to the data word. The receiver is instructed to expect either an odd number or even number, determined by the redundant bit.

For example:

Even parity	0	/	1	0	0	0	0	1	0
Odd parity	1	/	1	0	0	0	0	1	0

An error in the data word will change the parity, allowing the receiver to detect a transmission fault and request re-transmission. Adding a number of redundant bits can allow several errors to be detected and corrected. In some digital mobile systems used by the military, advanced error correction techniques have permitted the reconstruction of signals which have suffered bit error rates as high as 50 per cent, caused for example by jamming.

Most of today's error detection and correction techniques have been developed from the principles laid down by Hamming in the early 1950s. In Hamming's code, every seven digits contain four message digits and three 'correction digits', chosen so that the receiver can extract three equations solved to give a three-digit number. If the number is zero, there are no errors. If the number is other than zero, its value points to the digit received in error.

Consider, for example:

	a	b	c	d
4 bits of data	1	0	0	0

The check bits are:

Rule										
	e	=	1	+	0	+	0	=	1
	f	=	1	+	0	+	0	=	1
	g	=	1	+	0	+	0	=	1

The code word is therefore	a	b	c	d	e	f	g
	1	0	0	0	1	1	1

If 'b' is corrupted
The code word
received is   1   *1*   0   0   1   1   1

The check bits are used as follows:

Rule									
	e(1)	=	a(1)	+	b(1)	+	c(0)	=	2 = Fail
	f(1)	=	a(1)	+	b(1)	+	d(0)	=	2 = Fail
	g(1)	=	a(1)	+	c(o)	+	d(o)	=	1 = Pass

Therefore   a + c + d are OK

Therefore   b is identified as error

Data over radio

BCH coding (developed in 1960 by Bose, Chaudhuri and Hocquenghem), applies the principles of Hamming's code to provide multiple error correction.

Data over cellular

Cellular systems pose additional requirements for accurate transmission of data, to overcome transmission breaks (300 MS break at hand off cell to cell) and to avoid interruption from voice channel signalling.

Common methods include ARQ (automatic request re-transmission) and FEC (forward error correction and/or a combination of both methods. With FEC, up to half the data sent is for error correction. As bit error rates rise, falsing can occur, the codeword takes on a new legitimate, but incorrect meaning. The answer is to combine FEC and interleaving.

Data over cellular, particularly the data used for the signalling protocols, is often Manchester encoded. Ones are represented as 1/0 and zeros are represented as 0/1 to improve encode/decode accuracy at high bit speeds.

Shift keying

There are a number of ways in which digital information can be transferred to a carrier. Definitions of some of the common options are as follows:

FSK   Frequency shift keying. The modulation frequency shifts at the moment the bit stream changes its state. Logic 0 and logic 1 can be allocated to any frequencies between 300 and 4000 Hz.

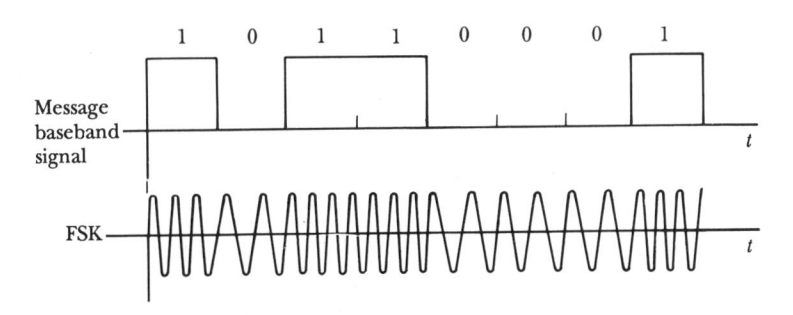

Figure 16.1 *FSK modulation*

MSK   Minimum shift keying. A variant of FSK with a frequency deviation equal to half the bit rate.
FFSK   Fast frequency shift keying. The phase at the moment of change is defined as the zero crossing point as a multiple of 180°. Transmission speed is normally 1200 bits/second.

Logic 0   three half cycles   1800 Hz
Logic 1   two half cycles   1200 Hz

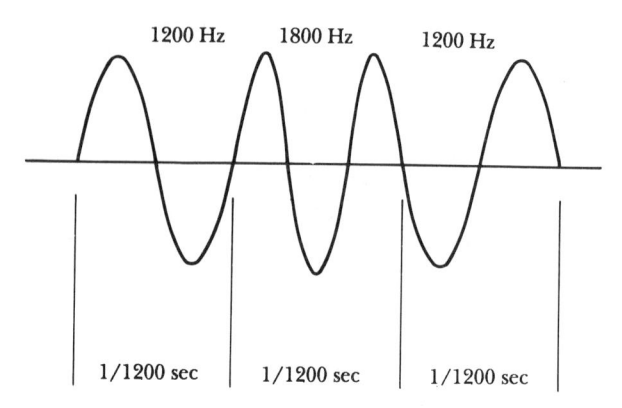

Figure 16.2 *FFSK modulation*

PSK   Phase shift keying (also known as signal phase space modulation). A 180° phase shift of the modulated signal at the zero crossing point. The reference phase has to be restored at the receiver.

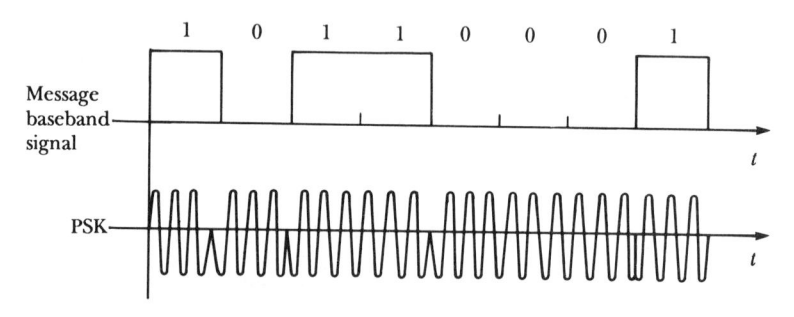

Figure 16.3 *PSK modulation*

Note FFSK and PSK have an advantage over FSK in that the moment of change is predefined, making it possible to use a coherent modulation/demodulation which is generally able to recover the data more accurately. The disadvantage of FFSK and PSK is that both systems are reliant on a stable phase shift relationship between transmitter and receiver.

QPSK   Quarternary phase shift keying. Two PSK signals running in parallel, 90° in phase.

DPSK   Differential phase shift keying. A phase shift corresponding to Logic 0 or Logic 1.

Dibit	Phase change
00	+45°
01	+135°
11	+225°
10	+315°

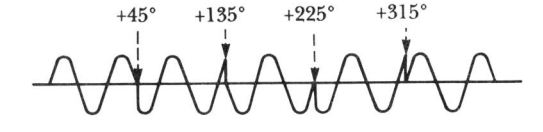

Figure 16.4 *QPSK modulation – four different phase changes allow two data bits to be sent with each change in phase*

Data	Phase change
0	+90°
1	+270°

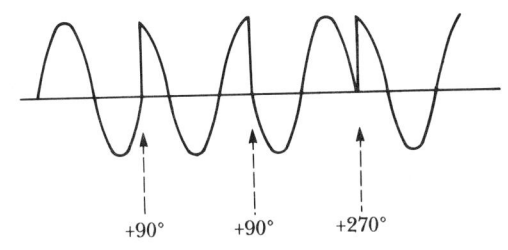

Figure 16.5 DPSK modulation

GMSK Gaussian minimum shift keying. The modulation technique used in the Pan-European GSM network. Similar to MSK but with additional pre-signal filtering (gaussian filters) to improve adjacent channel performance at a slight cost to intersymbol interference.

Bandwidth

GMSK is a 'tamed' digital modulation method (tamed FM is another example). 'Tamed' digital modulation techniques are used to achieve high data transmission rates – up to 16 kbps for a 25 kHz channel while still retaining adequate protection against adjacent channel interference.

Channel capacity

Channel capacity can be calculated by using Shannon's theorem in which

$$C = W \log 2 \left(1 + \frac{S}{N}\right)$$

where C = channel capacity in bits/sec
       W = bandwidth of the channel
       S = signal level
       N = noise level

Mobile Radio Servicing Handbook
Applications

Data messaging over radio has found particular favour in applications where ambient noise levels are high – for instance in fire tenders for the transmission of critical incident information.

Appendix I Abbreviations and glossary of terms

Abbreviations

RRD	Radio regulatory division (UK)
RIS	Radio investigation service (UK)
ERP	Effective radiated power
DTMF	Dual tone multi frequency
CTCSS	Continuous tone controlled signalling system
	Continuous tone controlled sub-audio squelch
dB	Decibel
PBX	Private branch exchange
AM	Amplitude modulation
FM	Frequency modulation
PM	Phase modulation
SSB	Single side band
SSBSC	Single side band suppressed carrier
RF	Radio frequency
IF	Intermediate frequency

Stations

Repeater	A transmitter/receiver communication used to extend the range of radio communication by relaying the signal
Unattended repeater	A remotely controlled repeater station
Trunked base	A number of base stations each working on a separate channel
Common base	A single channel base station shared by all users (also known as a community repeater)

Mobile Radio Servicing Handbook
Frequencies

Hertz	Cycle per second: the unit of frequency
KiloHertz (kHz)	1 thousand Hertz
MegaHertz (MHz)	1 million Hertz
GigaHertz (GHz)	1000 million Hertz
Audio band	The audible frequency range from 0.3–3 kHz
Microwave frequency	The frequency band greater than 1 GHz
Very high frequency (VHF)	Frequencies from 30–300 MHz
Ultra high frequency (UHF)	Frequencies from 300–1000 MHz

Appendix II Terminology

General terms:

Amplitude modulation	Changing the amplitude of the radio signal in proportion to the audio signal
Antenna	Any structure or device used to collect or radiate electromagnetic waves.
Antenna gain	An apparent increase in radiated power due to the design of the antenna
BALUN	Circuit used to connect balanced to unbalanced line
Circulator	A device used to separate signal paths in the upper VHF range of frequencies and above
Directional	Mainly in one direction
Doppler fading	Frequency related fading
Duplex operation	Simultaneous transmission in both directions
Effective radiated power (ERP)	The power measured at the output of the antenna in the direction of maximum gain
Encryption	The processing of the speech signal prior to transmission to prevent unauthorized eavesdropping.
	Common methods of encryption:
	Band splitting – change of voice frequencies to random sub-frequencies
	Time division – disassembly of message into separate time slots
	Frequency inversion – increase/decrease of frequencies juxtaposed
Frequency modulation	Changing the radio frequency in proportion to the audio signal
Frequency spectrum	The total range of (radio) frequencies
Intermodulation	Mixing and subsequent addition or subtraction of more than one frequency
Omnidirectional	All directions simultaneously
Phase modulation	Changing the phase of the radio frequency in proportion to the audio signal

Propagation	Emission of electromagnetic waves from an antenna
Radiating cable	A cable designed to radiate power along its length
Rayleigh fading	Distance related fading
Simplex operation	Transmission in only one direction at a time
Talk-back transmitter	A low powered transmitter which is added to a paging receiver to provide two-way communication
Unity gain	In an antenna, when the input power is approximately equal to the radiated power
VSWR	Voltage standing wave ratio, the 'match' between the antenna and its input line, with a maximum generally acceptable ratio of 1.5:1.

Component terms and circuit modules

AFC	Automatic frequency control
AGC	Automatic gain control
ALC	Automatic level control
Bi-polar transistors	A current controlled amplifying device
Ceramic resonators	Low cost synthetic mix used as resonant element in filter. Poor performance at > 10 MHz
Class A operation	The normal condition of operation for a single amplification device. It operates in a linear mode and indicates that the device is not biased off for any part of the cycle. High power requirement (due to inefficiency) but low distortion.
Class B operation	The condition of operation where the amplifying device is biased almost to the point of cut off. The amplifier is configured using a pair of devices where each only amplifies one half of the cycle. Used for power amplification due to higher efficiency.
Class C operation	The condition of operation where the 'amplifying' device is biased hard off and only part of the input waveform causes conduction. This results in pulses of harmonically rich output being 'switched' into a resonant network. Used for frequency multiplication.
Colpitts configuration	Oscillator configuration using capacitors in feedback path.

Appendix II

CMOS	Complementary metal oxide semiconductor
Dia electric	An insulating material separating two conducting surfaces.
Discrete devices	Individual semi conductor (active) devices. Bi-polar NPN/PNP J-Fets/P channel/N channel diodes
ECL	Emitter coupled logic
Electrolytic capacitor	A high capacitance type (>1 μFd) having a low impedance at tens of hertz.
Field effect transistor	A voltage controlled amplifying device.
Helical resonator	A high Q VHF/UHF inductor technique using a coil operating in a cavity
Hot carrier diode	A minority carrier diode formed by a metal to silicon junction
LSI circuit	Large scale integrated circuit, a device having 1000–100 000 active devices, typically performs a major circuit function (microprocessor or memory function).
MOSFET	Metal oxide semiconductor field effect transistor
NPN	A bi-polar transistor having the collector current consisting of electrons
OP AMPS	Operational amplifiers
Passive components	Inductors Capacitors Diodes
PIN diode	A silicon semi-conductor diode having pure resistance at RF. The resistance can be varied between 1 and 10 000 ohms by forward current control.
PNP	A bi-polar transistor having the collector current consisting of holes
Quadrature detector	Frequency modulation ratio detector
Resonant circuits	Circuits which discriminate strongly in favour of a particular frequency.
SAW	Surface acoustic wave
Shottky diode ring mixer	Frequency changer frequently used at UHF.
Strip line	A metal strip (e.g. pcb track) designed to have specific impedance at RF.
TTL	Transistor–transistor Logic
Varicap diode	A semi-conductor diode optimized for high capacity changes with reverse voltage.
VCO	Voltage controlled oscillator
VFO	Variable frequency oscillator

VLSI circuit — Very large scale integrated circuit, a device having usually in excess of 100 000 devices, typically performs a major system function.

Measurement terms

Current (I)	Rate of flow of electricity – amperes
Milliamp (mA)	Thousandth of an Amp
Microamp (μA)	Millionth of an Amp
Impedance (Z)	Restriction to flow of current, in ohms $I = \dfrac{E}{Z}$
Resistance (ohms) (R)	Real part of impedance
Reactance (ohms) (X)	Unreal (quadrature) part of impedance Reactance of inductor $= jwL$ capacitor $= \dfrac{1}{jwC}$
Farad (F)	Measurement of capacitance
μF	Millionth of a Farad (F $\times$ 10^{-6})
NF	Thousandth of a μF (F $\times$ 10^{-9})
pF	Millionth of a μF (F $\times$ 10^{-12})
Henry	Measurement of inductance
mH	Thousandth of a Henry (H $\times$ 10^{-3})
μH	Millionth of a Henry (H $\times$ 10^{-6})
Voltage (E)	Electrical potential
PEP	Peak Envelope Power
RMS	Root Mean Square – calculated by squaring all values, adding the squares together, dividing by the number of measurements (mean) and taking the square root of the result. RMS averaging gives an increase in measurement accuracy when noise is present.
Microvolts	A millionth of a volt (10^{-6})
Millivolts	A thousandth of a volt (10^{-3})
Watt	Power measurement Volts RMS $\times$ Amps RMS
Milliwatt	Thousandth of a Watt
Microwatt	Millionth of a Watt
dBm	dB relative to 1 milliwatt typically in 50 ohm system
dBd	Directional gain in dB (relative to a half wave dipole gain)
dBi	Directional gain (relative to an isotropic radiator for antenna gain)

Appendix II

Production terms:

Surface mount	The placing of components on, not through, the substrate.
Vapour phase soldering	Cloud of inert solvent – condensation of vapour on assembly heats by latent heat.
Wave soldering	*Dual Wave* Turbulent wave forces solder into joint areas Second wave clears away excess solder *Complex Wave* combines both turbulent and second wave High soldering temperature 260°C

Historical terms:

Cat's whisker	Early type of detector – two crystalline substances in contact with each other or a crystal in contact with metal, forming a conducting path whose resistance varies according to the direction and amplitude of the voltage applied.
Fleming's Thermionic Valve – 1904	A device passing current in one direction only, producing a DC output from a radio frequency AC input. *Application* – Applying DC current to a telephone receiver diaphragm to recreate an audible tone.
Lee de Forest's Triode Valve – 1907	Addition of a control grid to Fleming's design. *Application* – As a detector, amplifier and generator.

Appendix III The frequency spectrum

AUDIO SPECTRUM	RADIO SPECTRUM								
15 Hz–15000 Hz (15 kHz) Voice 300 Hz–3 kHz	Frequency	VLF/LF long wave 3 kHz–30 kHz	30 kHz–300 kHz	MF medium wave 300 kHz–3000 (3 MHz)	HF Short wave 3 MHz–30 MHz	VHF Very high frequency 30 mHz–300 MHz	UHF Ultra high frequency 300 MHz–3 GHz	SHF Micro-wave frequencies 3 GHz–30 GHz	30 GHz–300 GHz	LIGHT
	Wavelength	100 KM–10 KM	10 KM–1 KM	1000 M–100 M	100 M–10 M	10 M–1 M	1000 MM–100 MM	100 MM–10 MM	10 MM–1 MM	400–800 Nano-meters
		9 kHz Thunder-storm detection	30 kHz Coastal radio telegraph	300 kHz beacons	3 MHz Hearing aids distress	31.75 MHz Hospital pagers	300 MHz Radio positioning	3 GHz Radar	30 GHz Airborne radar	
		16 kHz Rugby Standard	60 kHz Rugby frequency standard	500 kHz distress	9.5 MHz BBC world service	35 MHz model aircraft	*425 MHz PMR UHF Band (17 major towns)*	Uplinks Downlinks	Aero-mobile	
		20 kHz frequency standard	160–180 kHz 280–315 kHz induction systems	526 kHz up broadcast medium wave	21–22 MHz fixed aeronautical	*41–48 MHz PMR (Band 1)*	*448 MHz PMR (London only)*	Fixed PTT 3.7 GHz 4.2 GHz 5.85 GHz	40 GHz Electronic news gathering	
		20–30 kHz coastal radio	150 kHz Long wave	1635–1792 kHz						

Commercial mobile communication frequencies

Frequency	Use
on 1/2/88 255 kHz	Aeronautical
	Space research
29.957 MHz	One-way pagers
29.96 MHz	Model control
76–78 MHz	lowband PMR
85–88 MHz	lowband PMR
88–97 MHz	lowband Radios 1, 2, 3, 4 and local radio
105–108 MHz	
138–140 MHz	PMR midband (ceases 1995)
139 & 148 MHz	Gas, electricity, coal
153	Wide area paging TX
163	Public radio phone
165	PMR High Band (12.5 KHz spacing)
175–216	
223–230 MHz	PMR (Band III)
614–854 MHz	
864–868 MHz	CT2 'Phone zone' 'Telepoint'
890–905 MHz	
935–950 MHz	TACS cellular
1000 MHz+	Satellite
1450–1525 MHz	Space and microwave
2900 MHz	Marine radar
830–915	
935–960	Pan-European Cellular (GSM)

Note Certain GSM frequencies have been allocated to TACS CELLULAR as an interim measure

Appendix IV Abbreviations for Mobile Radio Users

If channel loading is becoming a problem in an existing network, significant improvements in service level can often be achieved by reviewing speech communication disciplines on air, abbreviations and other short form references can be used to reduce average message lengths. Abbreviations recommended by the DTI are as follows:

ETA	Estimated time of arrival
ETD	Estimated time of departure
Wait or standby	Indicates that you are unable to reply immediately and is normally followed by an indication of time, i.e. wait/standby one means wait one minute
Say again	Report your last transmission
Roger	Your message has been received and understood
Over	I have finished my transmission and I await your reply. (This word is not used at the end of a final transmission.)
Please relocate	Means you are transmitting in a poor reception area, please move to a new area.

A	ALPHA		N	NOVEMBER
B	BRAVO		O	OSCAR
C	CHARLIE		P	PAPA
D	DELTA		Q	QUEBEC
E	ECHO		R	ROMEO
F	FOXTROT		S	SIERRA
G	GOLF		T	TANGO
H	HOTEL		U	UNIFORM
I	INDIA		V	VICTOR
J	JULIET		W	WHISKY
K	KILO		X	X-RAY
L	LIMA		Y	YANKEE
M	MIKE		Z	ZULU

Note, a recent survey by the Radio Regulatory Division of the DTI on a cross-section of London channels, indicated that within a typical busy hour only fifteen minutes of air time was being used efficiently for messaging, the balance being made up of repeat calls, or calls with no successful information transfer.

Appendix V   Private Mobile Radio Bands

Band	Type of service	Frequency band (MHz)	Total allocation (MHz)
Available throughout the United Kingdom			
VHF low All channels are at 12.5 kHz spacing	Mobile transmit (two frequency simplex)	71.5125–72.7875 76.9625–77.5000	4.025
	Base transmit (two frequency simplex)	85.0125–86.2875 86.9625–87.5000	
	Single frequency simplex	86.3000–86.7000	
Mid All channels are at 12.5 kHz spacing	Mobile transmit (two frequency simplex)	105.00625– 107.89375	5.875
	Base transmit (two frequency simplex)	138.00625– 140.99375	

Note, this band will cease to be used for Private Mobile Radio by 31/12/95. Additional spectrum will be made available in VHF Band I between 47 and 68 MHz

High All channels are at 12.5 kHz spacing	Base transmit (two frequency simplex)	165.050–168.250	7.350
	Mobile transmit (two frequency simlex)	169.850–173.050	
	Single frequency (simplex)	168.950–169.850	

Appendix V

Band	Type of Service	Frequency band (MH3)	Total Allocation in MHz
UHF All channels are at 12.5 kHz spacing	Base transmit (two frequency simplex)	453.025–453.975 456.000–456.975	3.850
	Mobile transmit (two frequency simplex)	£459.525–460.475 461.500–462.475	

Available in mainland of England, Scotland and Wales

Band	Type of Service	Frequency band	Total Allocation
VHF (Band 3)	Mobile transmit (two frequency) (simplex and duplex)	184.500–191.500 192.500–199.500 216.500–223.500	21.000
All channels are at 12.5 kHz spacing	Base transmit (two frequency) (simplex and duplex)	176.500–183.500 200.500–207.500 208.500–215.500	21.000

Available for use in London, Birmingham, Manchester, Liverpool, Aberdeen, Newcastle, Middlesbrough, Leeds, Bradford, Halifax, Sheffield, Nottingham, Derby, Leicester, Preston and Edinburgh and Glasgow areas

Band	Type of Service	Frequency band	Total Allocation
UHF† All channels are at 12.5 kHz spacing	Mobile transmit (two frequency) (simplex)	425.025–425.475 425.525–428.975	8.250
	Base transmit (two frequency) (simplex)	445.525–445.975 440.025–443.475	
	Single frequency (simplex)	446.025–446.475	

Available for use in London

Band	Type of Service	Frequency band	Total Allocation
UHF All channels are at 12.5 kHz spacing	Mobile transmit (two frequency) (simplex)	431.00625–431.99375	1.975
	Base transmit (two frequency) (simplex)	448.00625–448.99375	

† 12.5 kHz spacing is now being introduced into this band where previously allocations were at 25 kHz.

Total number of available channels for UK PMR is just under 1000 with 23 000 licensed base stations.

Note, the DTI generally consider that loadings of over 100 mobiles per channel are necessary to justify an exclusive channel allocation.

Allocation within 30–960 MHz

40% + broadcasting
Balance   13 MHz aeronautic
            18 MHz maritime
            276 MHz government
            152 MHz civilian mobile radio

Public radiophone bands

VHF	System 4	158.53125–159.9125	1.38125
12.5 kHz channel spacing	Two frequency duplex mobile transmit base transmit	163.0375–164.4250	1.3875
UHF	Cellular radio	890.0125–904.9875	14.975
25 kHz channel spacing	Two frequency duplex mobile transmit base transmit	935.0125–949.9875	14.975

Total = 106.04375 MHz

Application note

25 kHz FM	Deviation normally limited to 5 kHz
12.5 kHz FM	Deviation normally limited to 2.5 kHz
Cellular 25 kHz	Peak Deviation 9.5 kHz

CT/2 phone zone

864–868 MHz

Pan-European (GSM) digital cellular network

890–915 MHz
935–960 MHz

Appendix VI Conversion tables

Power – dBm – Voltage in 50 Ω
dBm = 10 log (P/P$_{ref}$)
where P$_{ref}$ = 1mW
V$_{rms}$ = $\sqrt{P \times 50}$

Voltage – dB/1μV
dB/1μV = 20 log (V$_{rms}$/V$_{ref}$)
where V$_{ref}$ = 1μV

Power	dBm	Voltage (RMS) in 50Ω
1 kW	+ 60	224 V
100 W	+ 50	70.7 V
10 W	+ 40	22.4 V
1 W	+ 30	7.07 V
100 mW	+ 20	2.23 V
10 mW	+ 10	707 mV
1 mW	0	224 mV
100 μW	− 10	70.7 mV
10 μW	− 20	22.4 mV
1 μW	− 30	7.07 mV
100 nW	− 40	2.23 mV
10 nW	− 50	707 μV
1 nW	− 60	224 μV
100 pW	− 70	70.7 μV
10 pW	− 80	22.4 μV
1 pW	− 90	7.07 μV
100 fW	−100	2.23 μV
10 fW	−110	0.71 μV
1 fW	−120	0.22 μV

Voltage (RMS)	dB/1μ
10.0 V	140
3.16 V	130
1.00 V	120
316 mV	110
100 mV	100
31.6 mV	90
10.0 mV	80
3.16 mV	70
1.00 mV	60
316 μV	50
100 μV	40
31.6 μV	30
10.0 μV	20
3.16 μV	10
2.00 μV	6
1.41 μV	3
1.00 μV	0
0.708 μV	−3
0.501 μV	−6
0.316 μV	−10
0.100 μV	−20

Mobile Radio Servicing Handbook
Decibel to ratio 0–19dB

| ___Ratio downwards (−)___ || dB | ___Ratio upwards (+)___ ||
Power	Voltage		Voltage	Power
1.0	1.0	0	1.000	1.000
0.977	0.989	0.1	1.012	1.023
0.955	0.977	0.2	1.023	1.047
0.933	0.966	0.3	1.035	1.072
0.912	0.955	0.4	1.047	1.096
0.891	0.944	0.5	1.059	1.122
0.871	0.933	0.6	1.072	1.148
0.851	0.923	0.7	1.084	1.175
0.832	0.912	0.8	1.096	1.202
0.813	0.902	0.9	1.109	1.230
0.794	0.891	1.0	1.122	1.259
0.759	0.871	1.2	1.148	1.318
0.724	0.851	1.4	1.175	1.380
0.692	0.832	1.6	1.202	1.445
0.661	0.813	1.8	1.230	1.514
0.631	0.794	2.0	1.259	1.585
0.603	0.776	2.2	1.288	1.660
0.575	0.759	2.4	1.318	1.738
0.550	0.741	2.6	1.349	1.820
0.525	0.724	2.8	1.380	1.905
0.501	0.708	3.0	1.413	1.995
0.447	0.668	3.5	1.496	2.239
0.398	0.631	4.0	1.585	2.512
0.355	0.596	4.5	1.679	2.818
0.316	0.562	5.0	1.778	3.162
0.282	0.531	5.5	1.884	3.548
0.251	0.501	6	1.995	3.981
0.200	0.447	7	2.239	5.012
0.159	0.398	8	2.512	6.310
0.126	0.355	9	2.818	7.943
0.100	0.316	10	3.162	10.00
0.0794	0.282	11	3.548	12.59
0.0631	0.251	12	3.981	15.85
0.0501	0.224	13	4.467	19.95
0.0398	0.200	14	5.012	25.12
0.0316	0.178	15	5.623	31.62
0.0251	0.159	16	6.310	39.81
0.0200	0.141	17	7.079	50.12
0.0159	0.126	18	7.943	63.10
0.0126	0.112	19	8.913	79.43

Appendix VI

Decibel to ratio 20–140dB

Ratio downwards (−)		dB	Ratio upwards (+)	
Power	Voltage		Voltage	Power
10.0×10^{-3}	100×10^{-3}	20	10.0	100
631×10^{-3}	79.4×10^{-3}	22	12.6	159
3.98×10^{-3}	63.1×10^{-3}	24	15.9	251
251×10^{-3}	50.1×10^{-3}	26	20.0	398
1.59×10^{-3}	39.8×10^{-3}	28	25.1	631
100×10^{-3}	31.6×10^{-3}	30	31.6	1.00×10^{3}
0.631×10^{-3}	25.1×10^{-3}	32	39.8	1.59×10^{3}
0.398×10^{-3}	20.0×10^{-3}	34	50.1	2.51×10^{3}
0.251×10^{-3}	15.9×10^{-3}	36	63.1	3.98×10^{3}
0.159×10^{-3}	12.6×10^{-3}	38	79.4	6.31×10^{3}
100×10^{-6}	10.0×10^{-3}	40	100	10.0×10^{3}
63.1×10^{-6}	7.94×10^{-3}	42	126	15.9×10^{3}
39.8×10^{-6}	6.31×10^{-3}	44	159	25.1×10^{3}
25.1×10^{-6}	5.01×10^{-3}	46	200	39.8×10^{3}
15.9×10^{-6}	3.98×10^{-3}	48	251	63.1×10^{3}
10.0×10^{-6}	3.16×10^{-3}	50	316	100×10^{3}
6.31×10^{-6}	2.51×10^{-3}	52	398	159×10^{3}
3.98×10^{-6}	200×10^{-3}	54	501	251×10^{3}
2.51×10^{-6}	1.59×10^{-3}	56	631	398×10^{3}
1.59×10^{-6}	1.26×10^{-3}	58	794	631×10^{3}
1.00×10^{-6}	1.00×10^{-3}	60	1.00×10^{3}	1.00×10^{6}
316×10^{-9}	562×10^{-6}	65	1.78×10^{3}	3.16×10^{6}
100×10^{-9}	316×10^{-6}	70	3.16×10^{3}	10.0×10^{6}
31.6×10^{-9}	178×10^{-6}	75	5.62×10^{3}	31.6×10^{6}
10.0×10^{-9}	100×10^{-6}	80	10.0×10^{3}	100×10^{6}
3.16×10^{-9}	56.2×10^{-6}	85	17.8×10^{3}	316×10^{6}
1.00×10^{-9}	31.6×10^{-9}	90	31.6×10^{3}	1.00×10^{9}
100×10^{-12}	10.0×10^{-6}	100	100×10^{3}	10.0×10^{9}
10.0×10^{-12}	3.16×10^{-6}	110	316×10^{3}	100×10^{9}
1.00×10^{-12}	1.00×10^{-6}	120	1.00×10^{6}	1.00×10^{12}
100×10^{-15}	316×10^{-9}	130	3.16×10^{6}	10.0×10^{12}
10.0×10^{-15}	100×10^{-9}	140	10.0×10^{6}	100×10^{12}

Note: In determining decibels for current or voltage ratios the currents (or voltages) being compared must be referred to the same value of impedance.

Mobile Radio Servicing Handbook
dBm conversion table (50 ohm System)

dBm	P mW	VRms V	dBm	P μW	VRms mV	dBm	P pW	VRms μV	dBm	P pW	VRms μV
+20	100.0	2.24	−20	10.00	22.4	−60	1000	224	−100	0.1000	2.24
+19	79.4	1.99	−21	7.94	19.9	−61	794	199	−101	0.0794	1.99
+18	63.1	1.78	−22	6.31	17.8	−62	631	178	−102	0.0631	1.78
+17	50.1	1.58	−23	5.01	15.8	−63	501	158	−103	0.0501	1.58
+16	39.8	1.41	−24	3.98	14.1	−64	398	141	−104	0.0398	1.41
+15	31.6	1.26	−25	3.16	12.6	−65	316	126	−105	0.0316	1.26
+14	25.1	1.12	−26	2.51	11.2	−66	251	112	−106	0.0251	1.12
+13	20.0	1.000	−27	2.00	10.0	−67	200	100	−107	0.0200	1.00
+12	15.8	0.890	−28	1.58	8.90	−68	158	89	−108	0.0158	0.890
+11	12.6	0.793	−29	1.26	7.93	−69	126	79.3	−109	0.0126	0.793
+10	10.0	0.707	−30	1.00	7.07	−70	100	70.7	−110	0.0100	0.707
+09	7.94	0.630	−31	0.794	6.30	−71	79.4	63.0	−111	0.00794	0.630
+08	6.31	0.562	−32	0.631	5.62	−72	63.1	56.2	−112	0.00631	0.562
+07	5.01	0.501	−33	0.501	5.01	−73	50.1	50.1	−113	0.00501	0.501
+06	3.98	0.446	−34	0.398	4.46	−74	39.8	44.6	−114	0.00398	0.446
+05	3.16	0.398	−35	0.316	3.98	−75	31.6	39.8	−115	0.00316	0.398
+04	2.51	0.354	−36	0.251	3.54	−76	25.1	35.4	−116	0.00251	0.354
+03	2.00	0.316	−37	0.200	3.16	−77	20.0	31.6	−117	0.00200	0.316
+02	1.58	0.282	−38	0.158	2.82	−78	15.8	28.2	−118	0.00158	0.282
+01	1.26	0.251	−39	0.126	2.51	−79	12.6	25.1	−119	0.00126	0.251
0	1.00	0.224	−40	0.100	2.24	−80	10.0	22.4	−120	0.00100	0.224
—	μW	mV	−41	0.0794	1.99	−81	7.94	19.9	−121	0.000794	0.199
−01	794	199	−42	0.0631	1.78	−82	6.31	17.8	−122	0.000631	0.178
−02	631	178	−43	0.0501	1.58	−83	5.01	15.8	−123	0.000501	0.158
−03	501	158	−44	0.0398	1.41	−84	3.98	14.1	−124	0.000398	0.141
−04	398	141	−45	0.0316	1.26	−85	3.16	12.6	−125	0.000316	0.126
−05	316	126	−46	0.0251	1.12	−86	2.51	11.2	−126	0.000251	0.112
−06	251	112	−47	0.0200	1.00	−87	2.00	10.0	−127	0.000200	0.100
−07	200	100	−48	0.0158	0.890	−88	1.58	8.90	−128	0.00158	0.0890
−08	158	89.0	−49	0.0126	0.793	−89	1.26	7.93	−129	0.000126	0.0793
−09	126	79.3	−50	0.0100	0.707	−90	1.00	7.07	−130	0.000100	0.0707
−10	100	70.7	−51	0.00794	0.630	−91	0.794	6.30	−131	0.0000794	0.0630
−11	79.4	63.0	−52	0.00631	0.562	−92	0.631	5.62	−132	0.0000631	0.0562
−12	63.1	56.2	−53	0.00501	0.501	−93	0.501	5.01	−133	0.0000501	0.0501
−13	50.1	50.1	−54	0.00398	0.446	−94	0.398	4.46	−134	0.0000398	0.0446
−14	39.8	44.6	−55	0.00316	0.398	−95	0.316	3.98	−135	0.0000316	0.0398
−15	31.6	39.8	−56	0.00251	0.354	−96	0.251	3.54	−136	0.0000251	0.0354
−16	25.1	35.4	−57	0.00200	0.316	−97	0.200	3.16	−137	0.0000200	0.0316
−17	20.0	31.6	−58	0.00158	0.282	−98	0.158	2.82	−138	0.0000158	0.284
−18	15.8	28.2	−59	0.00126	0.251	−99	0.126	2.51	−139	0.0000126	0.0251
−19	12.6	25.1							−140	0.0000100	0.0224

Appendix VII Mobile communications and ISDN (Integrated Services Digital Network)

Mobile Communications will be going digital in parallel with the world's telecommunications systems. In both cases, the introduction time scale is being determined (a) by component technology, and (b) by cost – a combination of component and infrastructure cost.

Digital communications – line and mobile – are projected to be cost competitive to analogue systems in the early to mid-1990s.

The enabling technology is dependent on the development of new components – logic circuits, fibre optics, opto electronics, broad band time switches, and new techniques of digital signal processing.

The application of the technology will provide the basis for the convergence of mobile communications, telecommunications, computing and broadcasting. An appreciation of the basics of ISDN gives an idea of how the convergence will occur.

The delivery medium for ISDN can be:

1 Hard copper or aluminium land lines as used today – still, in most cases, the most effective option for point to point speech and data transfer.
2 Optical fibre – more expensive, but with the advantage of immunity to electromagnetic interference, broad band capability and low error rate.
  The first application of optical fibre technology is as recent as 1977 – British Telecom's first optical fibre link between Hitchin and Stevenage. The recent commission of the TAT8 Trans Atlantic optical fibre cable is an example of the technology as applied today, a transmission medium working at 557 million bits per second. Each fibre pair in the cable carries 296.6 million bits per second, providing for 8,000 voice channels at 64,000 bits per second. Digital voice compression techniques shoehorn five two-way calls into a single voice channel, giving the capability of handling 40,000 voice circuits, with error rates down to 1 error per billion bits.
3 Radio links – including TV broadcast.
4 Satellites.

Telecommunications in the 1990s and beyond will almost certainly be based on a combination of all four options.

The networks will move gradually from analogue – including digital over analogue using modems, to fully digital.

Bandwidth efficiency will improve in parallel with improvements in speech and image compression and processing techniques. 'Conditional replenishment' is an example in which only areas of change are detected, coded and transmitted.

ISDN is already being implemented throughout a number of 'private' (closed user group) networks. ISDN will be implemented into the public networks in two stages. Initially, as narrow band, using 64 kilobit channels (known as the Q931 standard). This offers two access options – basic rate or primary rate. Primary rate gives high speed access for data transfer. Applications requiring broadband such as video conferencing, will be serviced through the use of multiple channels.

The second stage will be the introduction of a broad band network with 140 megabit channels. This requires further advances in switching technology beyond the technology that exists today.

Within the broad band network, intelligent multiplexers will be used to decide how best to package information which could include speech, high quality audio, data, video or high definition TV, to make optimum use of the available bandwidth and tariff structure. Bulk data transfer at night is an example. Delivery of high definition TV by telephone becomes a practical option.

Standards compatibility will be one of the most important factors in allowing communications convergence to take place. In computers, work continues on the OSI Standard (Open Systems Interconnection). **Note**: There are 9 million personal computers in Europe, all waiting to talk to one another.

In the first stages of ISDN, two of the most important standards to be finalized include:

1 The CCITT 144 kilobit 2B+D Standard – 2 × 64 kilobit B channels for voice and data, and a 16 kilobit D channel for signalling.
2 The fibre optic interface standard FDDI.

Co-ordination of standards is being undertaken by the European Telecommunications Standards Institute (ETSI) based in Nice, France.

It is, nevertheless, confidently predicted that there will be buoyant demand for protocol translation products to overcome differences of technical language and standards interpretation.

Telecommunicatioins will dominate our lives through the 1990s. Telecommunications investment will total £400–700 billion over the period, with investment and revenues together constituting over 7 per cent of Europe's Gross Domestic Product (EEC Statistics).

The challenge for mobile communications will be to ensure that the GSM digital network will be widely compatible with the new ISDN Standards. The confident prediction of the pundits is that by 1999, every one call in four will be mobile and (probably) digital.

Appendix VIII Sources of help and advice

Department of Trade and Industry
Radio Regulatory Division
Waterloo Bridge House
Waterloo Road
London SE1 8EA
01-215 7877

Electronic Engineering Association
Leicester House
8 Leicester Street
London WC2H 7BN
01-437 0678

Federation of Communication Services
Keswick House
207A Anerley Road
London SE20 8ER
01-778 5656

Health & Safety Executive
Baynards House
1 Chepstow Place
Westbourne Grove
London W2 4TF
01-229 3456

Institution of Electrical Engineers & Institution of Electronic and Radio Engineers
Savoy Place
London WC2R 0BS
01-240 1871

Mobile Radio Users Association
28 Nottingham Place
London W1M 3FD
01-400-1518

Mobile Radio Servicing Handbook
Radio Society of Great Britain (RSGB)
Cranbourne Road
Potters Bar
Hertfordshire EN6 3JE
0707-59015

Society of Electronic and Radio Technicians
57/61 Newington Causeway
London SE1 6BL
01-403 2351

Note: MPT Specification documents (as mentioned throughout the text) are available from the DTI (RRD) Division, Waterloo Bridge House, London

Index

Abbreviations, communication, 257, 266
Advice sources, 277–8
Aerials, *see* Antennas
Amplification, simple, 159
Amplifiers, RF, 158
Amplitude clipping, 54–5
Antennas, 40–1, 181
   colinear, 192–5
   directional, 183–200
   dish, 217–18
   installation, 207–11
   other types, 201–3
   specification, 203–6
   yagi, 183–7
ASCII system, 251
Audio output, 52
Audio processing, 53–5
Automatic gain control (AGC), 95

Band III network, 20, 24, 233–4
Base stations, 176–9
Batteries, 223–4
BCH coding, 253
Beam tilt, antenna, 195
Broadcasting, 180

Capacitors, 145, 162–3
Cellular radio, 21, 237–40
Channel capacity, 14–15, 247–8, 255
   coding, 245
Clipping, 54–5
Coaxial cables, 209
Coders, speech, 9
Colpitts oscillator, 133, 157
Components:
   chip, 145, 166
   passive, 145, 161–3

   special, 147–8
Connector systems, 210
Conversion tables, 271–4
Crystals, quartz, 56–7, 63, 146–7
   ageing, 223
CTCSS tone, 108–9, 178, 229, 230–1
CT2 telephone applications
   (telepoint), 24, 248

Data telegram, 239
Data transmission, 7–8
Decibels, as measurement base, 93
Digital divider, 72–7
Digital techniques, 8–9, 97, 244, 248
Diodes, 147, 160
   variable capacitance (varicap), 65–6
Dipole, cardioid, 197
Distortion, intermodulation (IMD),
   86–8, 94
Dual modulus prescaling, 74–81
Dual tone multifrequency (DTMF), 229
Duplexers, 214–17
Dynamic compression, 53, 54
Dynamic range, 88

Earth curvature correction chart, 220
EEC Directive (1988), 180
Electromagnetic compatibility (EMC),
   173
Electromagnetic transmission (EM), 12,
   15, 173
Environmental problems, antenna, 212

Fast frequency shift keying (FFSK), 7,
   8, 230, 231
Fault finding:
   antennas, 211

Index

receivers, 125–37
transmitters, 137–44
Field effect transistors (FET), 146
　IGFET, 156
　JFET, 152–4
　MOSFET, 156
Filtering, 46–7, 68–74
Filters, duplex, 214–17
Foster-Seeley discriminator, 50
Frequency, 11–12, 258
Frequency allocation, 17–20
Frequency analysis, 80–90
Frequency definitions, 258
Frequency response, 91–2
Frequency spectrum chart, 264–5
Frequency synthesizer, 60

Gain, 88, 203–4, 205
Glossary, 257–63
Grounding, 213
Group special mobile (GSM) network, 244, 245

Health and safety, 212–14, 224
Help sources, 277–8
High frequency (HF) short wave systems, 6, 17

IF filtering, 46–8, 61, 105, 156
Impedance, 13
Inductors, 161–2
Insulated gate field effect transistors (IGFET), 156
Integrated circuits (IC), 5, 161
Integrated Services Digital Network (ISDN), 275–6
Interference:
　reducing, 176–9
　source of, 179–80
Intermodulation, 179, 210
　distortion (IMD), 86–8

Johnson noise, 85–6
Junction field effect transistors (JFET), 146, 152–4

Leach resistance, 170
Line signalling, 232

Link planning, 218–21
Loop filter, 68–70
　phase locked (PLL), 70–4

Maximum usable sensitivity (MUS), 97, 243
Medium frequency (MF), 14, 17
Mixing, 44–5, 95
Mobile communications, 275–6
Modulation, 7, 26–7, 246
　amplitude (AM), 7, 27–9, 32, 48, 49, 107
　cross, 43, 87–8, 95
　CTCSS, 108–9, 178, 229, 230
　frequency (FM), 7, 29–32, 49, 107
　phase (PM), 32–3
　pulse-code (PCM), 9, 33–4
Modulation purity checks, 94, 97, 108–9, 242
Modulator AF stages, 159
MOSFET, dual gate, 156
MPT DTI specification, 8, 207, 278

Networks, PSTN digital, 9
Noise, 84–5
NPN device, 152, 154

Oscillators, 57, 60, 63, 157

Paging, 22
Pan-European (GSM) digital cellular network, 244–9
Parity checks, 252
Phase comparator, 67
Piezo electric effect, 63
PNP device, 152, 154
'Polo mint' effect, 199–200
Printed circuit board (PCB), 165–7
Private mobile radio (PMR), 20, 23
　bands allocated, 268–70

Quieting, 104

Radiation hazards, 214
Radiation reduction, 173–5
Radiation resistance, 13
Radio detector, 50

Index

Radio frequency (RF) technology, 7, 26, 40–4, 56–9, 60, 85
   band allocation, 20, 264–5
Radio site, choice of, 206–7
Receiver tests, 97–105, 115–19
Receivers:
   fundamentals, 34–5
   parameter definition, 94–5
   superhet, 34–5
   tests, 97–105, 115–19
Regulators, high current, 158
Resistors, 146, 163
Resonators, 147

Satellite links, 19
Screening, 175
Selcall, 226–8, 231
Selective calling, 226–8
Selectivity, 94, 95
Semi-conductors:
   discrete, 146–7, 152–60
   handling, 164–6
   integrated, 147
Sensitivity, 95
   maximum usable (MUS), 84, 97
Shannon's theorem, 255
SHF, 17
Shift keying, 253–5
   fast frequency, 7, 8, 230, 231
Signal/noise distortion ratio (SINAD), 7, 84, 95, 97–104
Signalling systems, 226–8, 245, 247
   tests, 241–3
Single burst tone (SBT), 229
Single side band (SSB) system, 7, 29
Site selection, 206–7
Soldering, 166
Spectral purity checks, 107–8
Spectrum allocation, 20, 264–5
Spectrum analyser, 92, 93
Spectrum efficiency, 226
Spectrum propagation, 15–16
Spectrum utilization, 14

Speech synthesis, 244–6
Squelch (muting), 51–2
Standing waves, 12–13
Surface mount technology (SMT), 5, 167–71
Sweep rate, 105
System 4 radio, 21–2

Terminology, 259–63
Test calls, 222
Test equipment, 96–7, 111–14
Tests, 114–24, 222, 240–3
   receiver, 97–105
   transmitter, 105–10
Time and frequency analysis, 88–93
Track repairs, PCB, 167
Transistors, 146, 152–4, 156
Transmission techniques, Pan-European, 36–9, 246–8
Transmitters:
   design, 53–5
   parameter definition, 95–7
   tests, 105–10, 119–25
Trunking, 23, 232–7

UK TACS, 235, 238, 240

Variable capacitance diode (VCO), 65–6, 83
Very large scale integration (VLSI) devices, 33, 36, 249
VHF/UHF systems, 14, 17–18
VLF/LF systems, 6, 17
Voltage standing wave ratio (VSWR), 203, 205, 211

Wavelength, 11–12
   optical, 15
Workshop repair procedures, 5–6, 25, 171–2

Yagis, 183–7